Consumer-based New Product Development for the Food Industry

Consumer-based New Product Development for the Food Industry

Edited by

Sebastiano Porretta
Experimental Station for the Food Preserving Industry, Italy
Email: sebastiano.porretta@ssica.it

Howard Moskowitz
Mind Genomics Advisors, USA
Email: mjihrm@gmail.com

and

Attila Gere
Szent István University, Hungary
Email: gere.attila@etk.szie.hu

ROYAL SOCIETY OF CHEMISTRY

Print ISBN: 978-1-83916-139-1
PDF ISBN: 978-1-83916-333-3
EPUB ISBN: 978-1-83916-334-0

A catalogue record for this book is available from the British Library

The Royal Society of Chemistry is a charity, registered in England and Wales, Number 207890, and a company incorporated in England by Royal Charter (Registered No. RC000524), registered office: Burlington House, Piccadilly, London W1J 0BA, UK, Telephone: +44 (0) 20 7437 8656.

Visit our website at www.rsc.org/books

Printed in the United Kingdom by CPI Group (UK) Ltd, Croydon, CR0 4YY, UK

Editor Biographies

Sebastiano Porretta is senior researcher of the Italian Ministry of Economic Development (formerly Ministry of Industry) at the Experimental Station of the Food Preserving Industry, SSICA, Parma, Italy where is Head of the Marketing and Consumer Science Department. He is well known to the Scientific Community for research in the study of product development and quality. His original approach to studying the interactions between foods and consumers and evaluating the effects of ingredients and processing on food acceptability is considered groundbreaking. He is a professor of food processing and sensory and consumer science in many universities and author/editor of 30 books. Professor Porretta also serves on editorial boards of several international peer-reviewed journals. He is a member of the International Union of Food Science and Technology (IUFoST) Board of Directors (2020–2022), and fellow since 1998 of The International Academy of Food Science and Technology (IAFoST), an elected group of distinguished food scientists. Since 1997 he has been President of the Italian Association of Food Technology.

Consumer-based New Product Development for the Food Industry
Edited by Sebastiano Porretta, Howard Moskowitz and Attila Gere

Published by the Royal Society of Chemistry, www.rsc.org

Howard Moskowitz is a legendary product market researcher, experimental psychologist, and inventor of world-class market research technologies utilized by virtually every company that matters. In 2004, Dr Moskowitz was the subject of a New Yorker article by Malcolm Gladwell, 'the Ketchum Conundrum', which Gladwell followed up with a TED talk entitled 'Choice, Happiness, and Spaghetti Sauce'. Most recently, Dr Moskowitz founded Mind Genomics Associates in 2014 to investigate how people think about aspects of their daily lives in order to provide insights for business and social research purposes. Dr Moskowitz earned his PhD in Experimental Psychology from Harvard University. In a long and distinguished career in market research, he founded the *Chemical Senses* journal, the first scientific journal devoted to the study of taste and smell.

Attila Gere is a sensory scientist working at the Szent István University, Budapest, Hungary as a postdoctoral researcher of the Hungarian Academy of Sciences. He earned his BSc in food sciences (with a specialization on distilling) and his MSc in food quality and preference. He received his PhD in Sensory Analysis form Szent István University, Hungary in 2016. His thesis deals with consumer sensory evaluation methods and sensometrics. The thesis won the Karai János predoctoral award in 2015. He started his academic career at the Corvinus University of Budapest in 2015 and has been working as assistant professor at Szent István University since 2017. In the academic year 2016/2017, he won a postdoctoral scholarship awarded by the Hungarian Government to the most promising young talents. His major research topic is titled "Uncovering the subconscious consumer mind" and it covers multiple areas such as consumer sensory evaluation methods, eye-tracking, and sensometrics. He is a member of the Institute of Food Technologists and the Hungarian Chemical Society. He has been publishing research articles since 2014 and working in the Sensory Laboratory of SZIU. He established stable scientific cooperation with numerous research institutes around the world. He has published several research papers which received hundreds of citations resulting an h-index of 10 and a cumulative impact factor higher than 100. He has wide industrial connections as he has been working for international food companies. Currently, he is part of a research group which won the support of the Hungarian Scientific Research Fund. Attila lives in Budapest with his family, he enjoys hiking and reading.

Author Biographies

Gianluca Donadini holds an MSc degree in Biological Sciences (1994, University of Parma. Parma, Italy) and a one year postgraduate specialization in Oenology and Sommellerie (2006, Università Cattolica del Sacro Cuore, Piacenza, Italy). His research aims at investigating the complex interactions between consumers, the marketplace and industry. In detail, Gianluca's research aims at explaining why, when and how consumers will consume food, while simultaneously studying how food is produced and stored. This complex system is explored through different disciplines (economics, food science, sensory science, ethology, nutrition, psychology, sociology, marketing and anthropology) by means of a combination of qualitative and quantitative methods. After a major focus on food safety, risk perception and risk communications since the second half of the 1990s, his attention has gradually expanded to cover consumers' behavior, nutrition, diet and health, and most recently also sustainability and food authenticity. Beer and brewing is one of the focuses of his research activity that spans both large-scale brewed and craft-made beer. In addition, Gianluca's research focuses on pre-schoolers and aims at investigating patterns of food preference and acceptance in children.

Consumer-based New Product Development for the Food Industry
Edited by Sebastiano Porretta, Howard Moskowitz and Attila Gere

Published by the Royal Society of Chemistry, www.rsc.org

Somrita Gan Choudhuri (PhD, National University of Singapore) is an Independent Researcher based in Singapore. Her scholarship explores communication inequalities in contemporary society. She uses research methods to examine various communication processes, strategies, and tactics to explore access to communicative spaces that leads to communication inequalities for effective solutions. Currently, she is focusing on the digital landscape.

Somsubhra Gan Choudhuri is the co-founder and CEO of Ai Palette, a Singapore-based artificial intelligence startup. Using artificial intelligence (AI), predictive analytics and image recognition, Ai Palette helps its customers craft and launch new products by spotting and analyzing emerging consumer trends.

Somsubhra is a marketer by training, a salesperson by expertise and is passionate about product innovation, having worked closely with some of the world's largest fast-moving consumer goods (FMCG) brands for more than a decade. He has been listed as one of the Top50 Innovators at the 11th World Innovation Congress.

Ariola Harizi is a Business Management graduate. She has specialized in executive management and marketing, thanks to her managerial experience in different companies operating in Albania and lecturing at University. She is a PhD Candidate at the Slovak University of Agriculture, Slovakia.

Orkida Ilollari has participated in several conferences and seminars, and is the author of several articles and publications in scientific journals in Albania and abroad. She is the author of the book '*Macroprudential Policies – The Effect on Guaranteeing Financial and Banking Stability*'. Dr Ilollari has many years of experience in the banking sector in various managerial and decision-making roles. She currently holds the position of Director of Master Programs and is a lecturer in the finance department of the Faculty of Economics at the Mediterranean University of Albania.

Namrata Mehta-Shah Taking on entrepreneurial endeavors with Coup D'ePop and Xocolatti, an innovative chocolate company run by a husband-wife duo, Namrata wears the hat of Cofounder for both start-ups by assuming the role of Creative Head as well as Marketing Director. Being a health nut, adventurer, and hopeless romantic at heart drives her to pursue her passions as a culinary enthusiast.

Shradha Mehta Balancing roles as Co-Founder of Coup D'ePop and running Ecomm Sales and Strategy for Essilor-Luxottica, North America, Shradha thrives on innovative disrupters within the ever-evolving digital retail landscape. As an entrepreneur, she launched and sold a Fashion Tech company, Dress for the Day. A strong believer in positive vibes only, this off-the-chart extrovert starts every day with a go-getter attitude.

David Moskowitz is a data scientist who uses experimental designs to help Fortune 500 companies develop successful new products and services. He has a BS in MIS and an MA in Accounting. David currently works as the lead data analyst for CraneCo, a global manufacturing company. He is married to a wonderful wife and is blessed with six children.

Petraq Papajorgji, Canadian Institute of Technology, Editor in Chief: *International Journal of Agricultural and Environmental Information Systems* (IJAEIS). Professor Papajorgji has published a number of scientific papers and books.

Dalma Radványi is a food scientist who received her PhD summa cum laude in Analytical Chemistry from Szent István University, Hungary in 2017. Her thesis was part of an EU-funded project entitled 'Research of analytical measurement and computer processing of metabolic markers to develop the technology of compost production KMR 12-1-2012-0189'. She was awarded a Fellowship granted by the Republic of Hungary. She has been working as an assistant lecturer at Szent István University since 2016 and is currently a postdoctoral research fellow at the Hungarian Academy of Sciences. She is currently involved in an EU-funded research project entitled 'E-nose research center – analysis of natural odor patterns and their applications in agricultural sciences'. She has established an international reputation by presenting her results at several conferences around the world. Furthermore, she has established a strong relationship with agricultural companies who adopted methods she developed. She developed a detailed database of volatile organic compounds indicating the presence of harmful microbes present in mushroom compost. Additionally, she introduced a chemometric approach to distinguish raw chromatograms which speeds up analytical measurements in heavily time-dependent agricultural applications. She has published several research articles in highly respected journals in the fields of analytical chemistry and biochemistry, chemometrics and food and consumer sciences. Through tutoring activity, she has helped students win national scientific competitions. Dalma currently lives in Budapest with her family, she enjoys traveling, reading, and working on publications.

Robert Sherman is a Microsoft Certified Programmer with over 20 years of experience who specializes in web portals, web sites and SQL databases. He graduated from the University of Michigan in 1986, and runs an independent programming service, RobertSoft. He created the MProd360 system for rapid product optimization and the PVI360 (Personal Viewpoint Identifier) to extend the assignment of people to mindsets beyond the basic experiment.

Giorgia Spigno is Associate Professor of food science and technology at the DiSTAS – Department for Sustainable Food Process of the Faculty of Agriculture, Food and Environmental Sciences of the Università Cattolica del Sacro Cuore (UCSC). She holds an MSc degree in Food Science and Technology (1999, UCSC) and a PhD in Chemical and Process Engineering (2003, Università degli Studi di Genova, Italy). Since 2001 she has been teaching different courses in Food Processing, Food Processing Plants, and Food Packaging at both BSc and MSc degree level in Food Science and Technology and Food Technology and Sustainability at the Agrisystem PhD School (UCSC).

She has gained expertise in food processing, food packaging and plant design, optimisation, innovation and modelling, such as the characterisation and optimisation of food functional profiles and shelf-life.

Ardita Todri graduated in Banking & Finance at Padua University. Subsequently she has specialized in Financial & Risk Management, thanks to managerial experience in various international banking groups operating in Albania. She holds two PhD degrees, one awarded by the Polytechnic University of Lugano and another awarded by the Catholic University of Murcia.

Himanshu Upreti is the CTO and Cofounder at Ai Palette and leads their technical development and vision strategies. He comes with industrial experience in the background of big data analytics and advanced machine learning (ML). Prior to Ai Palette, he worked at Visa Data Labs building large-scale data platforms with ML and AI predictive modeling, where he successfully filed two trade secrets. He is an alumnus of IIT Guwahati with a specialisation in mathematics and computing.

Ryan Zemel A market researcher who assists entrepreneurs in identifying 'perfect' products and messages that resonate with target markets. He has improved the efficiency and success of the ideation stage for numerous products and brands. He has strong expertise in conjoint analysis and research coordination.

Preface

Do the authors finally find the magic formula to develop successful products?

No, they absolutely don't, so if a magic formula is what you are looking for please don't buy this book.

However, there is a great need for a book that describes and explains food development from the point of view of the consumer and away from the top down approach. With the new era of big data, more nuanced food development solutions are becoming available.

This is probably the first book to describe how this might happen. It combines research from the most radical thinkers in the food product development sphere.

Predicting the rapid acceptability of an innovative personal computer promises a rapid increase in technological consideration of its manufacturer. The projection of profits from a new line of health drinks suggests that the period of depression could end for the company, and its share value. Expected sales of a commodity or ingredient in anticipation of a corporate merger will indicate a significant financial return for investors.

These are just some of the effects of successful product development.

The ideas for new products come from a large number of sources: from the scientist in the laboratory, from the developer, from the irritated customer or from the innovation of the competitor.

Managers, convinced of the merit of their idea through interpretation keys and data sources, generate impressive amounts of forecasting and financial performance documents concerning the opportunities for a new product. Predictions of sales and profits for the next five years, together with the mixed atmosphere of challenge and success in the market-place, motivate them to organize a development team. As soon as the project is approved by the general management, things start to move. The different skill-groups

Consumer-based New Product Development for the Food Industry
Edited by Sebastiano Porretta, Howard Moskowitz and Attila Gere

Published by the Royal Society of Chemistry, www.rsc.org

involved begin to spend weeks discussing and almost never agree, particularly on responsibilities. Someone begins to argue against the project all the more strongly with the increasing success of the competitor.

Finally, a compromise seems to emerge between the marketing, R&D and production staff. Thus begins the period of maximum turbulence to try to make up for lost time, in which each figure involved sets unlikely deadlines for their 'suppliers', typically production, creatives, graphic designers and printers. This will calm down with the settling of economic trends and/or with the advent of a new idea; the degree of innovation can range from simple extension to the most extreme penetration (turning products).

In the first case, *i.e.* in the extension of existing products, the introduction of new features that generally meet specific market demands, often have the connotation of a defensive move aimed at keeping the consumer from switching to the competition.

The so-called 'me too' products, for example, are copies of successful competitors' analogues, in some cases with timid additional innovations, new to the company that introduces them, but not to the market.

So what is a new product?

In an absolute sense, it should be something that did not exist previously, while in a more relative sense, it can be something that has not yet been proven, that is perceived as new. In defining new products, the later, *i.e.* something that is perceived as new, is much more useful and profitable, especially for those who bet on it. A novelty can represent an opportunity, but also a problem; however, the concept of new in the relative sense can be perceived as new in the absolute sense by some.

A product is a multidimensional concept, with expectations/needs to be met and in which interest must be generated: it represents to an organization, a great opportunity to create value that will be translated in different ways by anyone who feels such interest, tangible or not, long- or short-term. Nevertheless, a successful product must necessarily have an intimate concept, on which everyone can agree, that is capable of supporting the promise of a sustainable advantage for the organization that created it.

A new product concept can be seen as a path from an essence to a specific level. This path can be schematized, in a completely arbitrary but useful way, on some level that will be gradually deepened in the course of the text and which will serve to define the critical points of development.

The text is written with friends and colleagues, opinion leaders in their areas of expertise, who, as proof of mutual esteem, have immediately responded positively to my request for collaboration on a particularly demanding work that deals with the topics with the necessary scientific rigor and application.

The following is a list of product development hot topics that are treated in the book:

- the actual sensory and consumer approach
- from focus group to mind genomics

- mind genomics as a model-driven development
- consumer testing with children
- artificial intelligence to identify ideas
- features for product labelling, packaging and advertising
- assigning people to empirically uncovered mind-set
- systematic and researcher proclivities in product design

The volume is aimed at scientists, technicians and operators; quality control/assurance, R&D, marketing and processing functions will find the whole book interesting although I do not expect them to be equally attracted to the different chapters. For this reason, each chapter can be consulted separately, with the personal hope that the reader can benefit from continuing to read even initially unplanned parts.

PS

While being sure that he too thinks the same of me, I have always considered Howard to be one of the most pleasant and capable people. I have known Howard since 1995, despite the over 6500 km that separate us and the fact that sometimes timetables prevent us from having a coffee together. Despite his having started with an 'advantage' we have aged by dint of writing books and articles to disseminate our approaches and results on how to develop new products. We have never kept in our pockets even a grain of sand of our knowledge from the reader.

I take this opportunity to tell him that for the next 30 years we will only publish comics.

Sebastiano Porretta

Contents

Consumer-based New Product Development for the Food Industry
Edited by Sebastiano Porretta, Howard Moskowitz and Attila Gere

Published by the Royal Society of Chemistry, www.rsc.org

Part One: Sensory and Consumer Based Development

CHAPTER 1

Food Development: The Sensory & Consumer Approach

SEBASTIANO PORRETTA

Head of Marketing and Consumer Area, Experimental Station for the Food Preserving Industry, Parma, Italy
Email: sebastiano.porretta@ssica.it

1.1 Why Sensory Analysis?

Too often in the past food companies have entrusted consumer studies to marketers, who traditionally have set the problem according to a comparative logic (product A is sweeter than product B) by performing, at best, innumerable tests of comparison in pairs. At the basis of this rather simplistic choice was the suspicion that a number of samples greater than two caused consumer fatigue or that this was the discriminatory limit.

In addition, marketing staff, by delegating agencies and interviewers for data collection, have also become aware of the fact that in order to reach the quota, *i.e.*, the critical number of consumers, they inevitably lose direct contact with their consumers.

The concept of quality for food products is in general very complex: the quality of a product is linked to the combined effect of various factors, such as the suitability of the raw material, the transformation and conservation processes, the elemental composition, the commercial presentation including the quality:price ratio and, to a considerable extent, the sensory properties.

The acceptability of a food product is largely influenced by its sensory properties, *i.e.*, by all those properties that affect above all the organs of taste, smell and vision (Figure 1.1).

Consumer-based New Product Development for the Food Industry
Edited by Sebastiano Porretta, Howard Moskowitz and Attila Gere

Published by the Royal Society of Chemistry, www.rsc.org

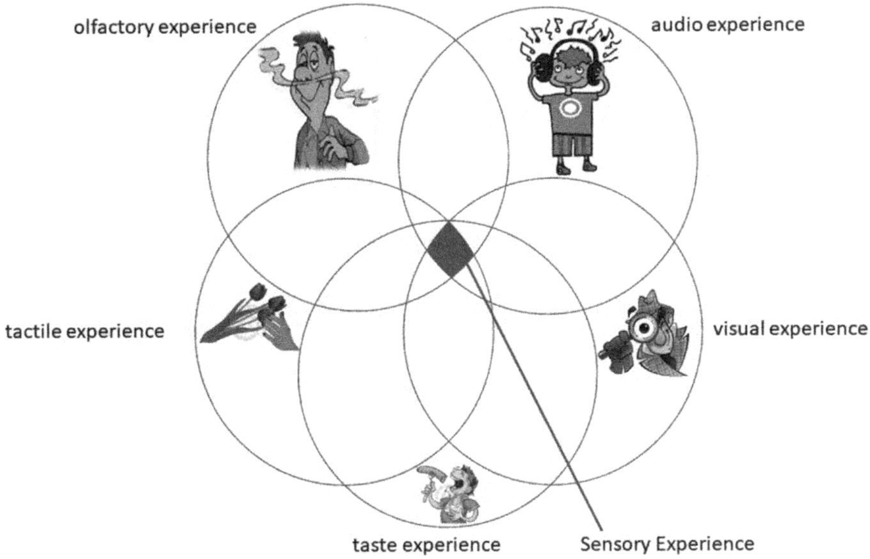

Figure 1.1 The sensory experience.

Many other aspects, such as nutritional value, chemical stability, the absence of pathogenic microorganisms, *etc.* are undoubtedly important in establishing the goodness of a food product, but the sensory characteristics give rise to the first judgment of the consumer, and a food that does not present a good taste, smell or appearance is unlikely to be appreciated. As mentioned, very often, especially in the past, the problems associated with sensory evaluation have been solved superficially; in fact, while for the determination of the physical and chemical properties, more and more precise tools and more and more specific techniques have been developed, the application of which requires careful preparation by the analyst, to evaluate the complex sensory properties, we have always thought that everyone could simply make a judgment. At the basis of this too simplistic behavior was the opinion that as people are different from each other, their sensory abilities are equally different, so that it is impossible to establish a judgment that is shared by all.

In food processing companies there is still a natural tendency on the part of both the research technician and the production technician to consider their personal preference with regard to flavor and smell an adequate guide for the development or improvement of a product. This self-confidence is often justified by the successes achieved, but in many cases said experts either did not work enough or did not work in conditions that offer convincing evidence of their reliability. In such cases the problem that arises is in determining the risk you take in accepting a given judgment, with all the economic and psychological inconveniences that may derive from it.

Sensory information is unique, and cannot be acquired by alternative means that would allow substantial direct costs advantages.

It is clear, however, that it is only in recent times that the commercial environment has shown a real interest in the potential inherent sensory evaluation and in how it can be best used. Such a recent, however slow, acceptance of this means of evaluation reflects the lack of a well-defined purpose and the inability of the sector experts to use their resources effectively. In fact, too much effort is spent on trivial questions, such as the search for a method that with a single test provides answers to all sensory questions, or in an attempt to educate people to respond as tools, capable of sensory responses closely related to specific physical and chemical values. Teaching people to provide a specific response assumes an invariability that is unrealistic and not typical of human behavior.

Sensory evaluation is a scientific discipline used to measure, analyze and interpret responses to products that are perceivable through the senses of taste, sight, smell, touch and hearing. The reader's attention is drawn to the fact that sensory evaluation is not limited to a specific sense, that it is therefore not possible to evaluate separately and individually the aroma or taste of a product, but that it is an inclusive process of perceptions which may lead to the identification of certain 'preferable' products, without guaranteeing a good result on the market.

Likewise, the use of sensory evaluation is not a guarantee of the success of a product. We know of companies with minimal or non-existent sensory resources and which are however very successful. There is no shortage of exceptions, which leads us not to put too much emphasis to these examples, but to go further.

The sensory characteristics of a product can generally be grouped into three categories, namely appearance, flavor and consistency. These categories, however, are not independent from each other. For example, color, which is obviously an important feature of appearance, can have an influence on flavor perception.

Consumers will assign higher scores for aroma intensity to dark red strawberry jams than light ones. The interaction between appearance and flavor is called 'visual flavor'.

Sensory analysis is a scientific discipline used to measure, evaluate and interpret the characteristics of food and other products perceived through the sensory receptors of sight, smell, taste, touch and hearing.

The definition goes beyond any qualitative and quantitative approach or assessments made by trained judges or with consumers. The use of appropriate statistical tools also ensures the reliability of the results. Sometimes improperly defined with the term 'organoleptic' to indicate the judgment provided through the sense organs, it is actually the sensory receptors and not the organs that respond to stimuli (Figure 1.2).

Examples of application of sensory analysis include the study of commercial life and shelf life, quality control and assurance, and the formulation of new products.

Sensory analysis was actually born to objectively describe what is highly subjective, *i.e.*, perception, and over time has significantly expanded its interpretative role to become a marketing tool, the latter which in its most technical sense refers to the study of the market (and not as it is interpreted in the communication/promotion of particular characteristics).

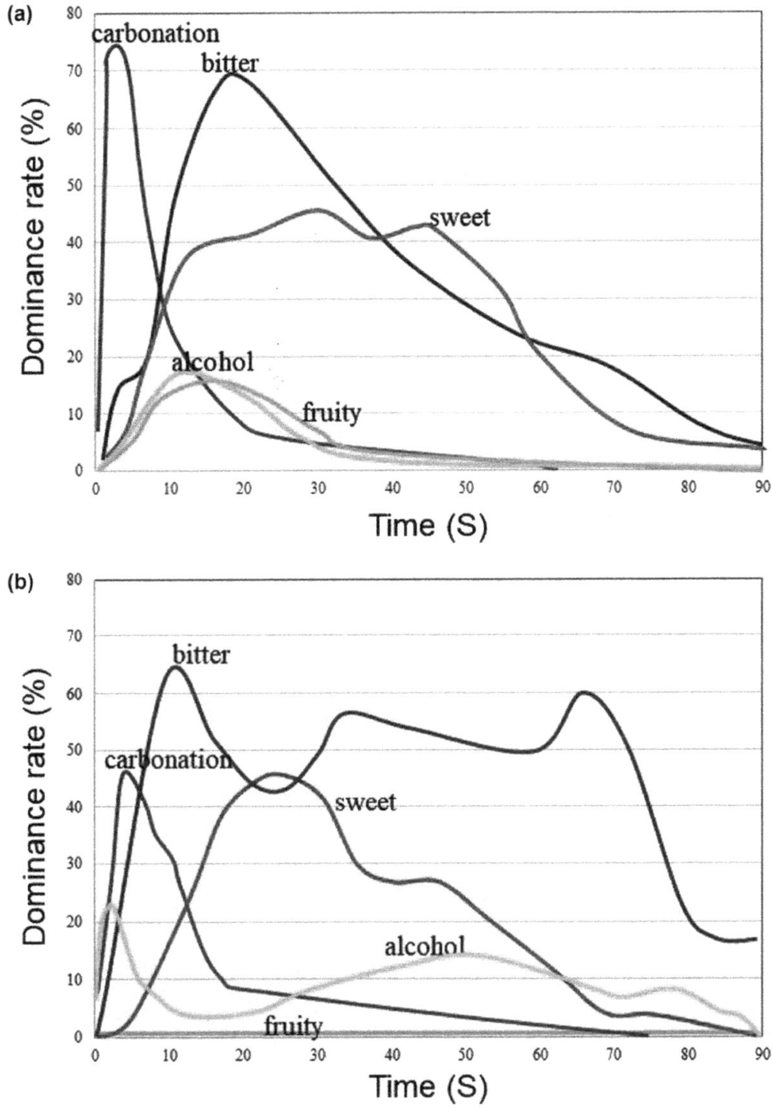

Figure 1.2 A dynamic time–intensity profile (a) and its change following an ingredient substitution (b).

1.2 Sensory Analysis and Its World of Myths and Legends

Sensory science has evolved into a pragmatic discipline perhaps even before developing its own intellectual corpus with a rational content and, as an unpleasant consequence of this, it reaches users and students burdened with considerable myths and beliefs about what the correct procedures are, what the objective is, and what is absolutely wrong.

Almost always, most of these legends have never actually been outlined and demonstrated in scientific literature through experimental data, but rather are shown/flaunted in meetings and in 'forums', especially by those who wish to take advantage of the elitist status that is thus generated. These are myths that are sometimes amusing, but which in fact represent a significant obstacle to the subject development, especially in its operation.

According to a more modern paradigm, the sensory analyst must be able to analyze the products, concepts and packages as a whole, while traditionally they have concentrated their studies on the products, leaving the research on the concepts to experts or market researchers. The sensory analyst must determine how the product fits the concept, as well as determine whether or not the product has the appropriate level of acceptability and the right sensory profile. The analyst must also check whether the efforts made by the entire team (R&D, marketing, production) meet consumer expectations or (reverse process), study the market expectations and communicate them to their colleagues in the aforementioned disciplines for a remodelling of the parameters studied.

Below, four legends are reported that still persist and are capable of delighting critics, and of perplexing neophytes, dogmatists and iconoclasts.[1]

(i) The consumer is unable to evaluate more than two products without losing sensitivity.

It is not clear at all where and how this sensory myth was born, because daily experience and psychophysics show that the average consumer is able to discriminate easily between a series of similar products and to correctly evaluate the sensory magnitude of different stimuli at relatively short time intervals, even in the absence of prolonged training periods.

The myth may have originated from studies in which each subject's goal was to discriminate between pairs of stimuli, rather than evaluating them individually on a scale. With N stimuli there are $N\ (N-1)/2$ pairs to be evaluated and if N is sufficiently small (*e.g.*, 4) the comparison generates 6 combinations of pairwise comparisons. When N is larger, the number of couples to be evaluated grows considerably making the goal impossible. For example, with $N=10$ a series of unmanageable pair comparisons is produced (45 pair combinations equal to 90 products to be evaluated), so it is more feasible to have the individual elements evaluated one at a time. Another possible origin of the aforementioned myth could be from impromptu comments expressed in company's kitchens, by 'assessment' or 'tasting' participants, or in any other product presentation setting. These are informal assessments of a number of products, often also competitors' ones. In general, these sessions do not require scientific rigor, more often than not, the company group arranges itself around the kitchen counter and opens the samples, inspects them and then tastes them. The sensory analysis expert presents the different samples in a predetermined order, *e.g.*, to show the specific 'nuances' between the products, then suggests a series of evaluation criteria, which can simply derive from the experience gained on the product or, more formally through the use of assessment forms.

Often when there is the simultaneous presence of technical staff and marketing experts, the order degenerates into chaos when the subjects examine one product after another, mixing them several times to try to better understand the differences. This chaos leads to negative comments, *e.g.*, by those subjects who do not 'remember' the product they have just tasted, who feel they have 'lost' their sensitivity or who do not remember much of the first sample in the series. These protests derive from the lack of pauses between the samples; in fact the participants are only allowed to rinse their mouths between one tasting and the next and are expected to keep in mind all the sensory information acquired. As a consequence, the participants feel they cannot cope with this sensory overload and complain that they cannot reach the goal. At this point then the careful researcher who attends the tasting session changes the procedure and allows each participant to take note of their feelings, thus simulating faithfully enough what happens in a real sensory test.

(ii) Sensory science to be conducted under optimal conditions must make use of strictly controlled conditions.

Sixty years ago few gave importance to the 'state of the art' and to the boundary conditions, while from a certain point on it was decided that the food sensory evaluation should be carried out only in separate booths in which the participants couldn't see each other.

A conspicuous number of these well-insulated booths with appropriate light and spittoons began to appear between the 1960s and 1970s; sensory analysis experts publicly boasted of this equipment as it offered the best ability to control the setting and, above all, it made the technicians involved belong to an increasingly restricted elite.

It was therefore decided that the only way to conduct any sensory evaluation was through the white booths, in which the judges spent the entire duration of the test in isolation.

If it is absolutely true that studies carried out under controlled conditions lead to the best results, more generally, what is needed for the judges is a comfortable environment, so that they do not suffer from the experience and the isolation imposed, especially in long sessions when evaluating numerous products, but are emotionally involved.

The extension of the methods of sensory analysis to consumer science, which corresponds to the analytical passage from the trained judge accustomed to the 'extreme' conditions of the laboratory to the subject/consumer, providing greater convenience, has made the requirement for isolation less important.

One way to pursue this dual objective may be to create an evaluation environment that includes computerized workstations for data acquisition and tables arranged so that the judges can see each other, without interacting and without seeing what the other judges have recorded. This provision is made by spacing the judges 'in the workstations' arranged radially so as not to perceive isolation, but that there is at the same time the required control. A limit to this provision may come from the privacy required when it is necessary to expectorate, in this case publicly, the products evaluated.

(iii) Only expert judges can evaluate the sensory characteristics of food, while the consumer judges only overall liking.

This is a particularly dangerous and well-established legend that has no justification in the scientific literature.

Some of its origins are based on scientific data, others on the desire to perpetuate a scientific practice, perhaps as a result of less noble and more elitist motivations.

Academic research, especially in psychophysics, began with the assumption that people could not function effectively as measuring instruments for assessing sensory intensity.[2] This is the so-called 'direct approach', rejected for more than three quarters of a century in favor of 'indirect measures'; the indirect measure presents the judges with two stimuli at a time and determines the degree to which they can be confused with each other or the level of preference. The subsequent statistical analysis evaluates the collected data variability and estimates the scalar values associated with the stimulus, with the property that the more two products are superimposable (or equally preferred), the more the stimuli will be found adjacent on the evaluation scale. Fechner himself, the founder of psychophysics, believed that the judges were not able to carefully evaluate the intensity, based his theory on the aggregation of units of 'discrimination' or 'minimally perceptible differences', JND (just noticeable difference), to build a psychophysical scale of magnitude. Thurstone developed a series of mathematical transformations to use this type of data on confusability by further expanding the approach, but always using indirect measures.[3]

Even the conviction about the impossibility of an evaluator to pay attention to multiple attributes of a stimulus have contributed to the spread of the myth. Psychophysicists did not say anything about the individual's ability to assess intensity, in fact the fundamentals of direct 'scaling' are precisely based on the assumption that panel members can judge intensity in a reproducible way through the use of scales. In the psychophysics labs of 50 years ago, academics involved in direct scaling instructed the judges to evaluate in one session only one particular aspect of the stimulus, *e.g.*, the generic body of a sound instead of treble, bass or loudness.[4] Even Stevens, one of the greatest modern psychophysicists, had a firm conviction that judges could not easily split attribute from attribute in a session.[5]

In the early scaling work on the sweetness and pleasantness of sugar and sweeteners, tests were organized so that panel members participated in two evaluation sessions. In one session they had to evaluate the sweetness perceived in the sugar and the other the pleasantness. Psychophysics scholars working with gustatory sensory stimuli capable of producing multiple sensations (sweet, salty, sour, bitter) wanted to free themselves from these limited and stereotyped paradigms and encouraged panel members to classify the level of each attribute on a scale. The result was that even those subjects not particularly trained did not seem to show problems in shifting attention from one sensation to another.

The commercial purposes of sensory science have contributed to the spread of the aforementioned myth. The 1950s and 1970s saw the interest in sensory characteristics shift towards descriptive analysis, and it was through some of these systems that the legend that the consumer was unable to evaluate a product's sensory attributes effectively arose, despite the findings reported by the scientific literature. All the descriptive methods require exhausting training; training was essential for panel members to understand the meaning of the attributes and to homogenise the terms meanings by using reference standards. Everything was fine until one had to deal with uncommon and unclear attributes without an explanation.

At this point much of the scientific research was stifled and collided with the statements regarding the inability of the consumer to evaluate the intensity perceived in a valid way on a scale.

It seems that it has never been established by the literature whether experts actually work better than non-experts, and there is not even a clear formulation of the criteria on which the experts are supposed to judge whether the experts are more sensitive to differences of the same stimulus. Do they have a better vocabulary? Are they more reliable? The results of the comparison conducted on different attributes between consumers and experts suggest that the two groups can be correlated in the classification carried out by operating with scales, when the consumer fully understands the meaning of the attribute.

(iv) Statisticians are the essence of a good sensory analysis (*i.e.*, 'If you don't know the latest statistics 'vagaries' you cannot consider yourself a professional').

This is a potentially harmful myth, in part because the abandonment of statistics could plunge sensory analysis back to the dawn. In part, however, the myth should be clarified, because for last 30 years statistical knowledge has been equated to professional competence in the field of sensory science.

Although the author of this (non-statistical) article is passionate about statistical analysis, especially the most recent one relating to the exploration and graphic representation of data, it often happens that those involved in sensory analysis use indecipherable statistical methods only to increase its prestige.

The data processing methods that stimulate the researcher to derive the greatest possible variability by using multidimensional modelling and a graphical representation of the data often blur the objective. Too often, in fact, concrete and valid thoughts are sacrificed in the name of modern statistical processing procedures, as if those who dealt with sensory science could not think about the problem without an esoteric approach. What often escapes the scholars of sensory analysis, a condition imposed on themselves by their own self-elevation, is pragmatism, *i.e.*, the possibility that the proposed method may be of real help (with a truly tangible use) in the development and evaluation of R&D and marketing products without the need for superior statistical knowledge, *i.e.*, far from promoting the use of elementary methods at all costs, there is an 'easy and ready to use' method even for the most complicated mathematical models.

It is necessary to dispel the latent belief, however, without effecting the impact of the sensory evaluation, that in order to obtain valuable insights it is necessary to employ complex statistical processing procedures.

Why do sensory analysis experts continue to accept mythology about consumer performance and correct operation? Scientists who are not sufficiently prepared, as has been the case for some time, are trained with myths, as the latter provide comforting directives and reduce operational uncertainty by providing a fake set of requirements. The sensory analysis expert captured by mythology does not need to ask questions, think or worry. Indeed, the myth prescribes a behavior and in that the neophyte scientist feels he is doing 'the right action'.

Furthermore, the myth creates a community. A community of professionals who work following the same mythology authorizes the neophyte to join it and the myth then becomes the entry price and the key to becoming a member.

Sensory evaluation is often the strategic key to determine the food products shelf life. The commercial life of microbiologically stable foods, such as biscuits or mayonnaise, is necessarily based on variations in their sensory properties.

In the development process, at the end of the supposed shelf-life the product concerned should be (a) similar to the standard/fresh product (just produced): this is typical in the case of a high-end commercial product for which the consumer is not prepared to accept any loss of quality when adhering to the storage conditions suggested by the manufacturer; (b) different, but acceptable, especially when it comes to natural products, such as cheeses or cold cuts, which are expected to be different over time (maturation/seasoning). The approach in these cases considers the following points: determination of the sensory profile for a series of products with different storage times to understand the changing profile mechanism over time; study of market acceptability; and integration of the aforementioned data to identify critical attributes.

1.3 Consumer Preference Segmentation

At the beginning researchers began in earnest to study what people liked and disliked. The researchers realized quite quickly that a lot of the basic likes and dislikes came from taste and smell. It was hard, in fact almost a forced 'task' to rate liking and disliking of texture and appearance.

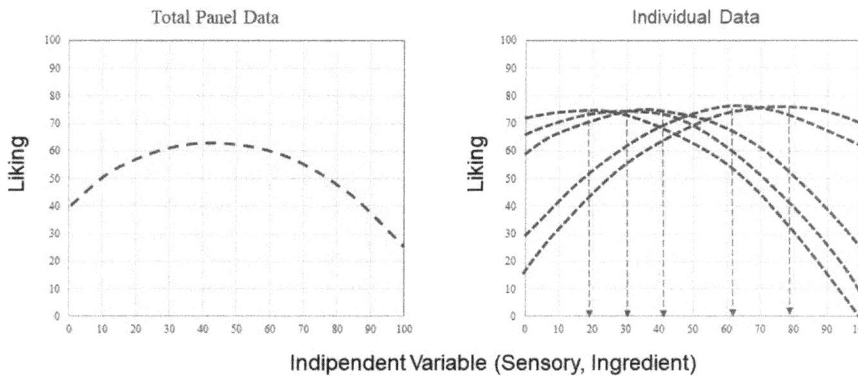

Figure 1.3 How overall liking changes as a stimulus changes in intensity.

It was thought then perhaps there might be some universally liked stimuli. One of these was oil of rose. It turned out that a lot of people liked this, but not everyone. The bottom line was that no stimulus in taste and smell was universally liked. There was always some degree of variability. That there is variability is one thing. But how does one go from variability to underlying basic genomes of the mind, basic patterns of likes and dislikes.

Look at Figure 1.3. The figure shows how liking changes as a stimulus changes in intensity. It really doesn't matter what the stimulus is. It's a matter of how the pattern looks.

The key to the whole problem is called an 'organizing principle'. The way Figure 1.3 presents itself, we are stuck with the measurement of inter-personal variation. It is nice to know that people differ from each other. But, truthfully, so what? Why would that interest us?

Let us move away from the world of cataloguing inter-personal variation and try to create an organizing principle. There won't be right or wrong; we are working on new ground, on *terra incognita.*

One way is to look for a simple organizing principle that will shed light. That principle is that we will look for the value of X, the sensory intensity, at which a person's liking reaches its peak, while at the same time ensuring that that sensory level (the optimal level), lies within the range of levels that we tested. In simple words, in Figure 1.3 (right panel), each person has a value on the x-axis where his liking peaks. These are shown by the vertical bars, dropping from the optimal to the abscissa, and defining a location on the abscissa corresponding to the optimum.

Using this principle to identify new segments and thus new products

Up to now we have talked in theory. Let us see how companies use this way of segmenting people, and then develop products. The approach is fairly simple. It is a matter of seeing patterns in these people-to-people differences and then using those patterns.

To describe the process we will go through the steps. At the end of the day, however, the proof is in the market. Does this approach really work? We will deal with that as well.

1. Relate liking on the ordinate (y-axis) to sensory intensity on the x-axis (abscissa). We saw an example of that in Figure 1.3.
2. Develop this same liking *versus* sensory relation on a person by person basis. Use the group sensory ratings for a set of related products, *e.g.*, orange juice, so that each product has one number for each sensory attribute. However, each person can use his own scale of liking.
3. The results for a group of individuals are going to look a lot like the different curves in Figure 1.3, but they will not be as nicely shaped. Some of the peaks will be higher, some of the peaks will be lower on the ordinate.
4. At the end of the day we will have a group of different numbers for each sensory attribute. Each person will generate one number for each sensory attribute. That number will tell us the level on that sensory attribute where the person's liking peaks.
5. Although some people may love the product and give it high numbers, and others may hate the product and give it low numbers, what is really important is the sensory level at which a person's liking peaks – no matter how high or low that level is.
6. We use statistical methods to divide people into groups, based upon the patterns of these numbers. Each person has a pattern of numbers to show the sensory levels at which their liking peaks. We can compare the people on these levels. Clustering methods, putting people into groups based upon the pattern of these numbers, produces new groups of people. People in the same group, or now the same 'sensory segment', show the same pattern of what they like.
7. This method has produced products in different forms and flavors, to satisfy the sensory segments.
8. To get a sense of just how powerful this method of sensory segmentation is, take a look at Figure 1.4. The data comes from orange juice, tested in three European countries. What is important is not the exact experiment, but what happens when our respondents test orange juice by amounts of pulp. In the three countries (left panel) we see similar patterns. But when we divide these respondents, not by country, but by sensory segment independent of country, we see different patterns, and the possibilities for different products (right panel).

Everyday experience tells us that people differ from each other in what they like. Observing people trying new foods, especially those with unusual flavors, quickly reveals to us the degree of this difference. And if we are still not convinced we need only go to the store, and watch people reach for different flavors of the same products. The person to person variability is unmistakeable.

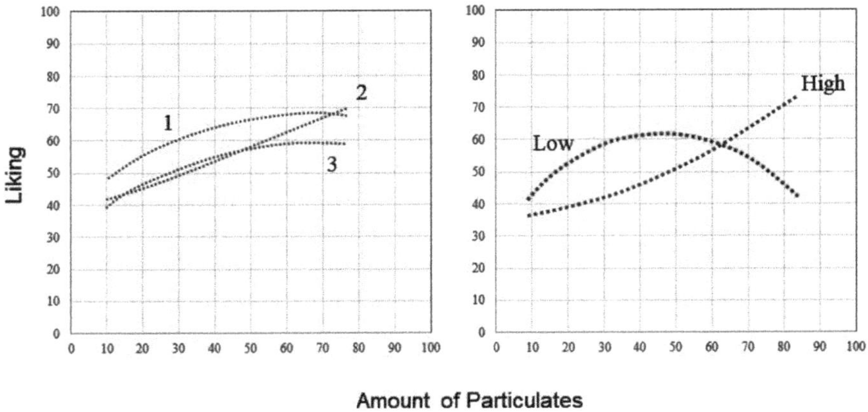

Figure 1.4 Sensory perception by country (left) and by panel (right).

A little deeper thinking and we ought to realize that Nature may present us with random variability, but not all the variation that we see in Nature can be attributed to this randomness. Maybe the variability is doing something else, signaling us that we are dealing with different groups of people who like different products. Then once when we emerge from looking for rules that apply to everyone, and start to think about different products for different folks, this once annoying variability, this intractable error terms, turns into our friend and our light, to help us see these different groups of sensory preferences scattered about in our target population. At the end of the day there lie new continents, new inhabitants, new products, rather than simply variability in what we already know.

1.4 Sensory Marketing

Sensory marketing deals with measuring and explaining consumer emotions in order to capitalize on them in new market opportunities. It also allows us to maximize the profit of particular products and customer loyalty so that they will repeat purchase so as to guarantee success for as long as possible.

Perhaps in an era characterized by short messages and rapid 'chirps', by further synthesizing we could define sensory marketing as those marketing techniques that deal with seducing the consumer through the senses. Sensory marketing deals with the upstream part of seduction, the study and measurement of what fascinates the consumer through his senses.

A sensory approach to market research tries to reduce the limits of traditional marketing which is too rational, based on the idea that a consumer is logically rigorous in his choices when faced with a product offer, competing products and so on.

Marketing has entered the era in which ideas and concepts must be reconsidered and in the same way their forms studied. However, it has set aside the need–purchase–benefit logic by creating new strategies capable, first, of studying the context and, therefore, of establishing emotional ties with consumers.

Mass marketing is definitively disappearing, replaced by new hyper-segmented markets in which individualism and personalization represent its strategic key. It follows that the necessary 'costs to improve' to reach consumers must be reviewed, supported by the numerous new forms of media communication that have appeared in the meantime. In the same way those tools, even the most recent ones, which have been used to obtain an express link with the market, including the CRM (Customer Relationship Management) for the management of customer relations or the CSM (Customer Standardization Management) for the specific marketing of customers, must be more personalized.

1.5 Actual Strategies in Product Development

There is big business in business advice, and there are some interesting patterns. There is the focus on process. Everyone is interested in process. Today's business is a process, a series of connected steps, of things one has to do. Then there is the focus on people. Everyone has to be given the chance to be all he can be. And finally, there is the focus on doing, on creating products, on doing more with less, on satisfying and evening delighting the customer.

Now that we have a sense of the value of process, as an organizing system that lets people marshal resources and get things done, we move to the world of process in product development. During the past 50 years or so, many companies have instituted product development processes. Their reasoning is quite simple and attractive. During the middle of the 20th century a lot of smaller companies were able to create foods, distribute them locally, and eventually put them into a wide number of stores. It wasn't easy, of course, but there wasn't as much competition and formalization as today.

With increasing sophistication and industrial 'savvy' comes the need to move the business beyond the talented individual, and into a system. The same can be said for product development. Whereas it was possible perhaps in 1950 or 1960 to create a product, put it out locally, see how it performed, and then modify it, over time the creation, selling-in, and logistical supply chain of products has become far tougher. Stores are not happy to have products to sell; they have valuable real estate to rent out to manufacturers. Stores do not just accept food virtually 'on consignment,' but rather rent out their space in terms of so-called slotting fees. So the story turns from the legendary single creator of a food to a food business. In short, the pattern is the same for food products as for just about everything else today.

With the risks involved, then, just how is the company going to succeed? It is one thing to have the genius product developer, or a recipe from one's grandmother. Those happen all the time; everyone has a grandmother somewhere or other, and a lot of the grandmothers were good cooks and made delicious products. But that is not enough. How does the company create a system that is independent of a grandmother or a genius developer? Most people do not have the heirloom recipe or the flair and genius to whip up a winning product. The answer is process.

When we think of those times, we do not think of a structured new products process for food. Indeed, reading stories about the history of foods and beverage products, as *e.g.*, in company's histories put out for public relations, we don't get a sense of process at all. The focus was entirely on the product, and some of the human interest stories attached to the product. Where there was process, that process was not formal. We get a feeling of the developer as an 'artist' or the developer as a normal housewife, creating the product, and somehow getting to stores, getting distribution, and growing the business.

By the 1970s, however, the world of management consultants had grown, and with that growth came the need to sell services. One of the most important of these services was 'process'. At one level process really could help the company become streamlined and more efficient. At another level, however, designing and then implementing the process created a revenue stream for the management consultant, who could bill hours for 'doing', and not just bill hours for thinking and advising. It is always a better deal for a company to be paid for actions that take a while (weeks, months) than to be paid for simple consulting that is done in a day or two.

So we have the development of processes for new product development. Many of these processes emerged from the efforts of management consultants, rather than emerging organically from the efforts of corporations themselves to become more efficient and effective. That is, the processes were introduced into the company as basic structures, on which the company's own processes could grow in a designed direction. It is somewhat like the introduction of basic lattice structures on a broken bone, around which the cells can regenerate in the proper form.

People do not buy foods or eat foods 'in a vacuum'. There is always some context, some situation. For many years marketers who identified opportunities for new foods have recognized that they must go beyond the normal methods to discover what consumers really want. When tracking studies became popular among consumer researchers in the 1960s it was all the rage to ask consumers about what more they wanted. The researcher could fold in these wishes to the tracking study, and present new opportunities to the marketer.

At the same time, however, qualitative researchers were asking people about what they wanted. Whether in focus groups or in-depth surveys, qualitative researchers were asking people to look inside themselves. These methods produced new ideas.

We now move from researcher-directed to consumer-directed. We reach the approach of the observation of behavior 'in context'.

1.6 Sensory and Marketing

The histories of sensory analysis and marketing research show two groups of professionals, both competing for territory, both focusing on one area, product testing, both bringing to the contest unique weapons. The sensory professional, schooled in science, comes from the discipline of knowing the product. The person who evaluates the product is of secondary importance.

What is primary is the product. One can almost call the subject or respondent the 'bio-assay'. The exceptional focus on the product can be seen in the way that sensory professionals think about the problems they address in their tests. First and foremost is whether the test answers the technical problem, and gives guidance. The other factors, such as venue, nature of the respondent, even nature of the questions, are, of course, important, but they don't play the same role as played by the solution to a technical problem.

The market researcher comes to the contest with a different set of rules, a different intellectual history, and different needs. No one expects the market researcher to be an expert in the product itself. Rather, the market researcher positions himself as an expert on insights about the respondent, the consumer, the marketplace. Whether the respondent is testing a product or completing an attitude and usage questionnaire about the product, the market researcher brings the same set of skills to the study. Indeed, it is rare for market researchers to be able to point to any particular skill in product testing. It is the general world of consumers, not the particular world of the physical product, that interests the market researcher.

In years gone by, from ~1970 to ~2000, sensory and market research were content to operate separately. They rarely had to deal with each other because, in reality, they occupied different worlds. Sensory and marketing research might clash from time to time, but the clashes were restricted to consumer evaluation. The sensory professional wanted to move beyond expert panels and small, localized 'guidance tests', to larger consumer tests. When it comes to consumer tests, however, all too often the market researcher perceived this desire for growth by the sensory professional as a 'land grab', rather than as an earnest desire to expand one's horizons.

The rather different worlds inhabited by sensory and market research were occasionally noticed and remarked upon, more often by those in R&D, rarely by anyone in marketing.

During the end of last century, the sensory world kept to itself, dutifully 'profiling' products using the experts that were trained on the different test methods. No isolation lasts forever. As the 1990s rolled on, competition increased. The sensory practitioner was soon being asked to provide more guidance to the product developer, not so much from the expert panel, as from the consumer. It was at this time that the first real clashes occurred with the marketing researchers, who, up to then, were doing the large-scale tests with consumers. Up to the 1990s market researchers were content to work with sensory professionals. There was a tacit understanding between the two groups that the sensory professional might, on some occasions, work with consumers, but the work would be of very small scale, perhaps involving local consumers recruited in the neighborhood of the R&D lab that housed the sensory researchers. In any event, the consumer research that sensory researchers might do would be of the type called research guidance: small scale panels, ~50 people, in the local area, to guide the product developer. In no way would the sensory person get into the product testing business in the way that the market researcher was.

Tacit agreements, a *modus vivendi*, and limiting one's reach, all have ways of disappearing over time. The corporation of the 1990s was facing different challenges from the same corporation 20 years before. In the 1970s, supporting an infrastructure of sensory professionals doing profiling work was acceptable, and even noteworthy when one could point to a spanking-new laboratory to house the sensory professional and his activities. The 1970s were still squarely in the wake of the recovery from World War II, when prices were stable, housing affordable, and competitors in the marketplace present, but not particularly threatening. By the 1990s that stability had gone away, and the corporations were fighting for customers, and often times for their lives. The world of the sensory professional and the world of the market researcher would change.

1.7 Competition Between Corporate Functions

By the mid-1990s the 20 year silence between the sensory world and market research world was coming to an end. As the pace of product development quickened, companies began to move more quickly. The time for running large-scale descriptive studies of products to get their 'footprint' was coming to an end. Of course, the changes were not announced as such. Rather, the status quo was changing. It simply was not the norm to do extensive descriptive tests. The field of product testing was moving on. Sensory was changing, and in that change it would butt up against market research, as sensory sought new areas in which to function.

By the mid-1990s it was becoming clear that rapid consumer testing would become the more important part of the sensory professional's job, and that the business would demand that the sensory efforts should move towards a practical and measurable outcome. It was no longer sufficient to provide product developers with reams of output showing descriptive analysis tables and even 'maps' of products' descriptive terms. Rather, the goal of sensory professionals was to provide actionable information to the company.

Disagreement and even open strife between the sensory department and the market research department was not long in coming. As sensory professionals began to position themselves as the low-cost supplier of product testing, they came into direct conflict with the market research department. Market researchers were accustomed to farming out, *i.e.*, subcontracting, the research of products to external agencies. In order to survive and compete, the sensory department began to position itself in the corporation as the 'lower cost supplier'. The sensory department was no longer merely the guardian of knowledge about the product from the sensory point of view, but rather was simultaneously a data provider with consumers.

Over time the positions of sensory professional and market researcher would be sorted out, leaving in its wake anger and suspicion of each other. To the market researcher the sensory department was an internal function, behaving like an interloper, usurping the job of market research, compromising and endangering the position of the market researcher in the company. To the sensory professional it was survival, the search for new areas, and the search for job security in an environment that had moved past the halcyon days of the 1970s and 1980s.

The sensory world is beginning to recognize a need to come together with market research. Some of the recognition is structural. Within the past 5 years or so, a great number of sensory professionals have lost their jobs. Whereas in years gone by the sensory professional could join a company, having been actively recruited, and look forward to a long and secure employment, that security is over and done with. Sensory professionals are as likely to be fired as anyone else in the corporation. There is no more comfortable, long-term employment in that field.

How then does this rapprochement happen, at least what are some early warning signals, because the coming together if it is happening, is just starting. No longer is the sensory professional talking about a seat at the table for the sensory professional, the need to be recognized, the need for its own corporate 'lebensraum', to quote a German war phrase. Rather, the talk in these meetings is about the practicalities of co-existence, about how some companies are 'doing it right, now', *i.e.*, having sensory and marketing research work together. There is less and less talk about me *versus* them, about the conflict of sensory and marketing research, about who should win. There is talk about how the two professional groups must work together to create a good future.

What then is this future?

1. Sensory professionals will continue to do a great deal of their traditional work in descriptive analysis, but will not talk very much about that 'bread and butter' work. Descriptive analysis has lost a great deal of its luster. It is service work now, a way to survive, but certainly not the future. Furthermore, no one much cares about descriptive analysis in terms of forward-looking initiatives; it is more of a clerical job, rather than something perceived to move the business forward. At a practical level, this lowered value of descriptive is forcing the sensory professional to seek some other means to survive and win job security.
2. Sensory professionals will 'lock up' the corporate role of testing products created by R&D. Being close to the R&D world and having the ear of the product developer provides the sensory world with a very strong position here. Rather than going to market research for this early testing among consumers, for it is market research that traditionally heard the consumer's voice, it is now the sensory professional in the company or on contract that will have this responsibility. Market research 'lost' that battle, or perhaps more diplomatically, ceded that ground.
3. Sensory professionals are looking for, and will continue to be looking for, new ways to understand the consumer. These new ways include ethnography, the observation of consumers 'in context', and biometric measurement of consumer responses to foods. In these new areas, also being investigated by market research, we may see a coming together of

the two fields. Neither sensory nor market research has sufficient experience in these new areas to claim them as part of the professional 'patrimony'. With the corporation demanding greater returns on research, these new areas of understanding the consumer allow the two fields to work together in order to survive the rocky seas of competition that shake the corporation from time to time.

4. Finally, in about 10 years or so, as the cooperation between sensory and market research increases, and as both sensory and market research evolve towards 'insights' rather than just the testing function, we may expect to see a *de facto*, if not formal, merging of the two departments, primarily as a cost-saving move. There will be no need for two separate departments providing insights about the consumer, when a more efficient, cost effective, single department can do the job.

References

1. S. Porretta and H. R. Moskowitz, *Innovare con la consumer science*, Chiriotti editori, Italy, 2011.
2. G. T. Fechner, *Elemente der psychophysik*, Leipzig, Breitkopf und Härtel, 1860.
3. L. L. Thurstone, A law of comparative judgment, *Psychol. Rev.*, 1927, **34**, 273–286.
4. H. R. Moskowitz and S. Porretta e M Silcher, *Concept Research in Food Product Design and Development*, Blackwell Publishing, Ames, USA, 2005.
5. S. S. Stevens, *Psychophysics: An Introduction to Its Perceptual, Neural, and Social Prospects*, New York, Wiley, 1975.

CHAPTER 2

The Changed Paradigm of Consumer Science: From Focus Group to Mind Genomics

SEBASTIANO PORRETTA

Head of Marketing and Consumer Area, Experimental Station for the Food Preserving Industry, Parma, Italy
Email: sebastiano.porretta@ssica.it

2.1 From Qualitative to Quantitative in Concepts Research

Often in focus groups the reactions to concepts are probed in agonizing detail, forcing out points of view and reactions that are merely incidental to the concept itself. Some of this in depth analysis leads to new insights, and in the end the insight improves the concept. Some of the in depth analysis is simply self-indulgence, and in the end is an exercise without other purpose. In contrast, most marketers are less interested in the nuances of concepts, and instead focus on performance. If left to their own devices, these marketers would throw away all of the diagnostic information in the concept, and simply deal with the performance of the concept on the one key evaluative dimension such as percent of purchase intent (*i.e.*, percent of the respondents who say that they would probably or definitely purchase this product). The testing process requires a structure in which to operate.

There are two major forms of testing. One is qualitative testing, such as focus groups or in depth interviews, the other is quantitative testing.

Consumer-based New Product Development for the Food Industry
Edited by Sebastiano Porretta, Howard Moskowitz and Attila Gere

Published by the Royal Society of Chemistry, www.rsc.org

In qualitative methods for idea generation the focus group interview has been adopted as a major data-gathering technique by market researchers who are interested in the appeal of advertising strategies or in consumer product preferences. In many focus groups the experts sit behind one-way mirrors, or now webcams, and watch respondents. Sometimes experts sit with the consumers. Occasionally the process is inverted, and the consumers sit with the experts.

Focus groups, comprising small groups of individuals, or even in-depth interviews with one person, allow the free exchange of feelings and reactions to concepts. The interviewer, trained in eliciting information from the respondent, can probe the respondent about his reactions to concepts, and ask other members of the group to comment. The interaction can be very fruitful. Those involved with the concept can sit with the group members, or more frequently, as noted above, sit in a more comfortable viewing room, away from the respondents, and shielded by a one-way mirror. The concept writer, the marketers, the researchers listen to the dialog between the focus group moderator and the respondent, noting key ideas where they emerge. At the end of the focus group, or usually at the end of a series of groups, conducted in different markets, with different respondents, the moderator writes up his observations of reactions to the concepts, and other relevant information. Often the reactions provide a wealth of information about what the consumers like about the concept, and what they do not like. These reactions are often accompanied by suggestions about what specifically to change. Occasionally, and more frequently today than before, the moderator may use the focus group to obtain consumer ratings of the concepts. There are too few respondents, at least in the mind of a quantitative researcher, but the practice continues to occur.

Concepts are critical for the development and marketing of products and services since they constitute the structure for these products and services, albeit at the level of consumers rather than at the technical level. Market research practitioners and their corporate clients are well aware of the need to create winning concepts. To that end a very large practice has emerged in the business world designed to understand the current competitive environment, identify opportunities, and convert these opportunities into concepts. Look at any prospectus by a market research company and there immediately emerges a self-proclaimed expertise on concept research. In actuality, however, the scientific underpinnings of concept research are quite weak. There is a dearth of both practical and scientific information about how to create and evaluate concepts.

Although practitioners provide services for evaluation and optimization, the publications offered by the scientific community and the academic business researchers do not go into depth about how to create and measure concepts, leaving corporations and their developers/researchers in the lurch. This gap between practical needs and absence of solid scientific information can be traced to the fact that, until recently, both concepts and concept research have been relegated to the realm of application. There has been little focus on establishing relevant, scientifically accessible knowledge bases for concepts,

Table 2.1 A product concept *versus* a positioning concept for a soft drink.

Product concept
Introducing healthy drink
All the nutritional features of regular yogurt
With active cultures
A sugar-free beverage
Available in fruit, and tropical flavors
Positioning concept
Drink designed for good nutrition
Its active cultures improve digestion
Specially formulated to be low fat so you don't need to feel guilty
An array of flavors, plain, fruit, and tropical, specially selected for your sensory delight

although the scientific literature will attest to keen interest in the general topic of food product development. Concept development in the new product development literature is conventionally relegated to the so-called 'fuzzy front end'. Yet it is the concept that often directs product development and marketing. A good product concept can help make the product a success by guiding developers and advertising in the right direction.

The lack of information about creating good concepts is most pronounced at the corporate R&D function. Most R&D departments either work with the concepts that marketers provide, or are instructed to come up with a winning new concept and then, in turn, create a product to match the concept. For the most part those assigned to this job are poorly prepared, and have nowhere to turn.

Concepts comprise written statements about products or services. At first glance it seems intuitively easy to deal with concepts because they are similar to the stimuli that one has encountered over one's life: the written word applied to the description of a product or a service. To create the concept one might need to do some quick writing, jot down the features of a food or beverage, talk perhaps about the way the product is packaged, consumed, perhaps move on to a statement about health benefits, price, and somewhere within the description bring in emotional elements, or a tie-in with a famous person. Furthermore, there does not seem to be the technical barrier to research on concepts that pervades research on actual products. The investigator need not be a scientist; there are no esoteric formulas to master, no complex equipment to understand, and no formidable chemical or physical principles to comprehend. Certainly, concepts should be easy to understand, and concept development and research should be straightforward (Table 2.1).

2.2 If Boredom Exists, How Can We Measure It and Use It?

Can we quantify boredom? Can we predict which products will be most subject to hedonic habituation, so that the manufacturer can concentrate on those stimuli which will maintain their acceptance over a long time-span and pay less attention to those stimuli that will lose their attractiveness once they are widely used? Why are some fragrances popular year after year, while other fragrances (or other 'looks' in cosmetics) are popular for only a limited time?

Part of the answer to the problem of prediction comes from food science rather than from cosmetics and toiletries. Food researchers know that consumers like some foods more than they like others, and that some items (*e.g.*, hamburger) never seem to lose their attractiveness. Other foods that score well on food attitude surveys (*e.g.*, steak) quickly lose their attractiveness if the consumers have to eat the same product day after day. (In contrast, hamburger maintains much of its intrinsic acceptability if the consumer eats hamburger day after day.) Why? Are there any general rules or trends that could generalize across categories, and help the manufacturer of personal products become smarter in the selection of sensory characteristics?

2.2.1 Some Products Become Boring Faster Than Do Others

In food product evaluations we can ask consumers to rate their interest in consuming a food, given a variety of occasions during which they previously consumed the same food. (This is an attitudinal question, not a behavioral question. We ask the consumers to estimate how interested they would be in the product, assuming a previous length of time since last consumption.) Some simple foods like bread never achieve high levels of acceptance, but also never really become boring. Other foods, such as steak, becoming boring rather quickly, even though these items intrinsically score very highly.

Could it be that fragrances follow the same patterns? That is, some of the more complex yet identifiable fragrances can become very boring after repeated use, whereas other fragrances which are less identifiable, less immediately acceptable but also less 'salient' maintain their modest level of acceptance.

2.2.2 The Degree to Which a Product Is "Boring" Is Inversely Related to the Last Time the Product Was Selected and Consumed

The time preference curve suggests that interest in a product increases as a function of the time since the product was last consumed. This means that even the most boring product can increase its acceptance (however little) if the consumer stays away from the product. Does this hold for a fragrance? Does time since the fragrance was last used affect boredom or wearout, or is boredom with fragrance an 'all-or-none' phenomenon (*viz.* once the consumer becomes bored with the fragrance the fragrance never again regains its acceptance)? This is an area worth pursuing for basic and applied research. It may well turn out that fragrances follow a one-time preference curve (all-or-none – *viz.*, once boring then never again accepted), whereas products designed for more visual or tactile use show a more graded time preference curve (*viz.*, continuous – acceptance increases up to a maximum asymptotic level if the consumer does not use the product for a long period of time).

2.2.3 Asking the Proper Question for Time-preference

Time factor in product acceptance is a new area of research for which the scientific literature is not well-developed. Consequently there are no 'tried and true' procedures whereby the researcher can construct a questionnaire to accurately probe the time effect of repeated usage and boredom. One could ask simplistic questions such as, 'How likely are you to become bored with this product if you were to use it every day?' One could ask a more complicated question such as, 'How interested are you in using the product if the last time you used this type of product, with these same sensory characteristics, was *x* weeks ago?' The former question asks the consumer to assign an overall rating of estimated potential for 'boredom'. The latter question asks the consumer to hypothesize about interest in the product, given various rates of usage. By asking the panelist to rate liking as a function of time since last usage rather than as a function of actual rate of usage, the researcher 'concretizes' the problem of usage frequency. Panelists have an easier time responding to concrete questions rather than responding to more theoretical questions. Time since last usage is a concrete question. Frequency is a more theoretical, more abstract question.

2.2.4 The Concept of Time Preference

In the late 1970s, Joseph Balintfy and his colleagues began to investigate the phenomenon of changing food preferences.[5] Balintfy noticed that food acceptance could not be characterized by a single number. Rather, there seemed to be two components – one a basic level of acceptance, and the other an effect of time and repeated exposure to the food. According to Balintfy's theory (and subsequent confirmatory observations), the acceptance of a food product is not a static phenomenon, but rather a dynamic one. Foods are optimally acceptable when they have not been consumed for a very long time. However, just after consuming a food, an individual might not want to consume the food again. Acceptance drops as a function of recent eating. Over time the food regains its initial level of acceptance. Eventually, after a sufficiently long wait, the food becomes just as acceptable as it had been originally. Balintfy also noted that some foods are more 'time sensitive' than others. For instance, meats and complex dishes were more sensitive to 'time since last consumed' than were other foods (*e.g.*, breads). Although steak is highly acceptable, when the consumer has just eaten it acceptance plummets. It takes a while for the consumer to desire steak again. On the other hand, some products like milk and bread are not highly acceptable to begin with, but also never lose their acceptance, even on repeated consumption. These items seem to be time independent.

2.3 Trade-off Analysis

The initial applications of conjoint analyses used trade-off procedures. The respondent would be presented with two options, and instructed to select which of the two options was preferred. From the array of trade-off judgments made by the respondent the researcher would then create a scale comprising a utility value for each stimulus such that the pattern of choices would be recaptured if the respondents were to have used those utility values to make their pair-wise choices.

Trade-offs make historical sense when viewed against the background of psychometrics and marketing research, even if to us today they seem tortuous and indirect. Experimental psychologists in the latter part of the 19th century assumed that the human judge was incapable of acting as a measuring instrument. With this presupposition, therefore, it was left to the ingenuity of the newly emerging psychological scientist to devise a method for measurement that did not use the human judge as a measurement of magnitude. Gustav Theodor Fechner (1860), the founder of modern day psychophysics, averred that a better way and probably a more valid way to measure sensory magnitude was by analyzing the behavior of the judge who tried to compare two stimuli, either to make a direct match or to say which of the two was more intense.[1] One act of matching consisted of adjusting two stimuli so that their perceived magnitudes were equal. Fechner believed that one could measure the error of adjustment, and subsequently, through analysis, one could then convert the observed error to a measure of underlying subjective magnitude. Fechner's strategy, measurement error in the matching of two stimuli, would need an algorithm that converted measurement error to underlying psychological units. Another method was to have the panelist identify which of two stimuli appeared more intense, or was more acceptable, or in general had 'more' of a specific aspect. The degree to which people differed in their choice (*viz.*, a measure of error across people, or a measure of an individual's error across stimuli or replications, respectively) was assumed to have a relation to subjective magnitude. Two stimuli that were very close together in subjective magnitude were thought to be confused more frequently – hence the converse – the degree of confusability was assumed to parallel the similarity in subjective magnitude.

This paired comparison approach appears in Figure 2.1, which shows a screen-shot of two concepts for a soft drink. The two concepts each have a picture, and a set of descriptor terms. The respondent selects one of the two, and in some rates the magnitude of preference. From the selection patterns made to a set of such pairs, the researcher can construct a model showing the part-worth contribution of each of the elements. The methods for creating this type of utility value vary from processing the rank orders of the choice to regression modeling, and are described in texts devoted to conjoint analysis.

When dealing with trade-offs, it is important to keep in mind that the researcher assumes the respondent to be unable to act as a measuring instrument, *per se*, and thus unable to provide a direct rating of element utility. Rather, the respondent is assumed to be able to judge pairs of products, and to make a selection. In the history of experimental psychology this approach is called the indirect method. The 'indirect' refers to the fact that the respondent never really assesses utility *per se*. Rather, the utility is deduced from the choice patterns. The choice patterns need not be perfect. The indirect method recognizes that the choices of people are 'error-prone'. To the

Figure 2.1 Example of a choice study in conjoint measurement, where the respondent chooses between two stimuli. Utilities are constructed from the pattern of choices.

degree that the choice behavior is 'noisy', and ful of inconsistencies, the utility values of the different elements will be closer together. To the degree that the choice behavior is 'not noisy', and the patterns are cleaner, the utility values will be farther apart, and show a clear hierarchy.

It is worth noting that the indirect methods, exemplified by paired comparison and subsequent processing, continue to be well accepted by market researchers. Often when one hears the term 'conjoint analysis', one also hears the accompanying clarification that the method is 'trade-off'. A lot of people accept this approach. Indirect methods are often justified by the researcher's conviction that respondents really don't know the utility values of the components in concepts. Thus, in light of this argument, the researcher moves towards the use of trade-off methods.

The history of product testing followed the same path. Over the years researchers have used the method of paired comparisons for product testing, even though in the end the researcher had to use extensive *post hoc* statistical manipulation to bring the array of paired comparisons down to a single scale. The concentration of effort on the statistical machinery to process paired comparisons left little time for the researcher to do more with the utilities than estimate them, and perhaps put them in a model.

2.4 Full Conjoint Profile

An alternative way to deal with the problem of trading off requires that the respondent be presented with a set of combinations, already created by experimental design. The experimental design combines these components to generate complete concepts, but these complete concepts, in turn, comprise elements that appear independently of each other. That is, although the respondent may

Figure 2.2 Example of a single concept as a combination of features, and the rating question. In conjoint measurement the respondent would evaluate a variety of combinations.

perceive the concepts to be 'complete' or 'whole' ideas, similar to the way complete ideas appear in Nature, the experimental design ensures that the reaction to the whole can be deconstructed into the part-worth reactions. There are many different ways to combine elements into small, easily read combinations. Whichever way is chosen, however, the approach develops the combinations in a structured fashion amenable to subsequent regression modeling.

Figure 2.2 shows an example of the stimulus for this approach. The respondent need only look at one concept at a time, which simplifies the effort. The respondent can rate this concept on a scale, or simply choose to accept or to reject the concept. When the researcher combines the elements by experimental design, it becomes straightforward through subsequent analysis to estimate the contribution of each concept element to acceptance/rejection rating, or to the decision to select or reject, respectively. The conventional tools of multiple linear regression work fine for this analysis. The conjoint analysis will then use regression modeling to show the application of the method in different situations involving concepts.

2.5 A Soft Drink Development

The use of a soft drink as a product is particularly suitable for making several variations both in formulation and in packaging, required by an existing or presumed market.

Let us imagine that a company intends to develop (similarly, it would be worth reformulating or reviewing) a soft drink that covers the demands of a healthy, functional and sporty trend.

Table 2.2 Conceptual factors and relative levels.

Type	100% Juice 100% Juice + Vitamins A, C, E Juice and pulp + mineral salts Nectar + ginseng
Packaging	PET with push and pull sport cap Brik with drinking straw Glass Brik with a pour screw cap Can Maxi pack
Sport claim	Absent Present (reference to particular sport)
Health benefit claim	Absent Present (ingredient properties)
Natural claim	Absent Present (natural, organic)
Emotional effect	Refreshing Thirst quenching Strengthening Absent
Taste	Orange Apple Pear Peach Exotic mix

The study phases for the development of the new product will be:

(a) systematic study of concepts;
(b) creation of prototypes;
(c) sensory tests;
(d) product optimization;
(e) market tests.

Naturally, we will focus on the first point, leaving the other aspects to later articles.

First of all, it is necessary to establish the products, the conceptual factors and the relative levels to be examined (Table 2.2).

Data collection takes place, as mentioned, through questionnaires that display the concepts in the form of 'cards' with exemplifying images.

The images on the 'cards', together with any brief descriptions, have the task of simplifying the amount of work for each consumer, who normally evaluates about 20 conceptual combinations, and, at the same time, contain part of the information present in the experimental design.

An example 'card' is shown in Figure 2.3.

Figure 2.3 Example of a card with exemplifying images.

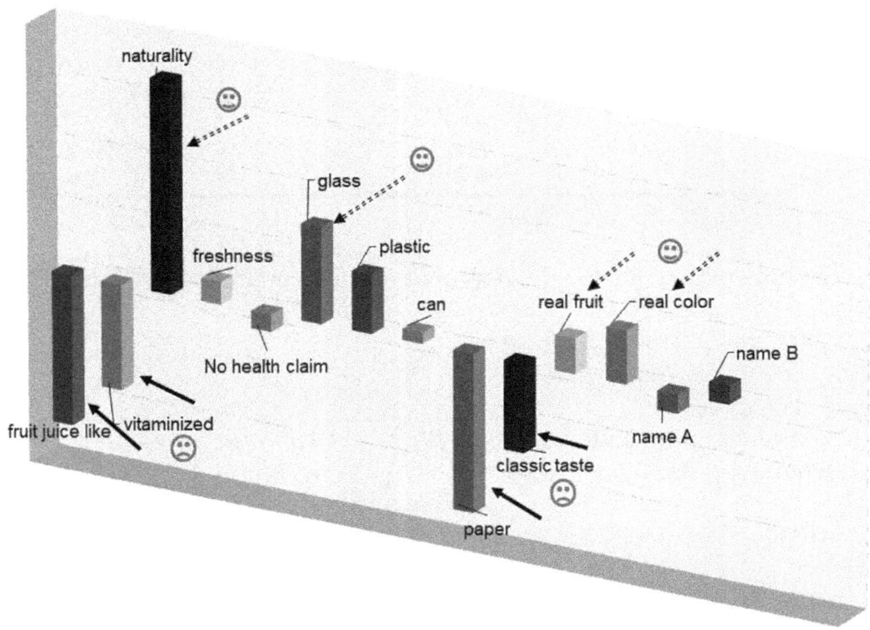

Figure 2.4 Utilities from conjoint analysis.

The analysis can be carried out globally (on the products) or individually (on the subjects), obtaining in the latter case the relative importance for each subject, while the utilities can be obtained for each combination of factors or for each factor.

The test results are shown in Figure 2.4.

The utilities that most positively guided the consumer's overall preference for the new drink were:

- 100% vitaminized juice A, C, E;
- juice and pulp with mineral salts;
- presence of a health claim;
- PET with push and pull sport cap;
- Brik with a pour screw cap;
- orange flavor;
- exotic mix taste;
- thirst-quenching;
  - invigorating;
  - presence of a sport claim;
  - absence of a natural claim;

while, in a negative way:

- nectar with added ginseng;
- Brik with straw;
- glass packaging;
- absence of an emotional claim;
- absence of a sport claim;
- presence of a natural claim.

It should be emphasized that the utilities reported derive exclusively from the processing carried out and, therefore, it is not at all consequential that if the absence of a claim entails a positive utility, its presence will lead to a negative one (it could be indifferent or even positive, perhaps less intense).

2.6 Market Segmentation

Groups defined by the use of the same brand or belonging to the same gender, age, social group, *etc.* represent the most typical way used by marketers to divide the entire population in question.

There is a more convenient way of dividing the population based on reaction to concepts. The original approach is known as segmentation.

Briefly, sensory segmentation assumes that there are groups of individuals that show reliable, albeit different, paths/ways towards the acceptability of a product. For example, one segment of people may like a floral fragrance more and another a vanilla one. The specific details determined in the case of the soft drink development are shown in Table 2.3.

The first segment is made up of consumers who consider the product simply as a new (in the sense of another) drink – refreshing, invigorating, 100% juice, indifferent to the packaging.

The other group (second segment) is made up of consumers who want a more trendy product, *i.e.*, oriented towards a sporting, healthy and natural culture.

Table 2.3 Market segmentation.

Category	Location	1st positive utility	2nd positive utility	3rd positive utility	1st negative utility	2nd negative utility	3rd negative utility
Male: 9–16 years	North	PET with push and pull sport cap	100% juice + A, C, E vitamins	Invigorating	Glass	Nectar + ginseng	Pear
	Centre	PET with push and pull sport cap	100% juice + A, C, E vitamins	Yes sport claim	Glass	Nectar + ginseng	Pear
	South	Brik with a pour screw cap	Juice and pulp + mineral salts	Yes sport claim	Nectar + ginseng	Glass	Pear
Female: 9–16 years	North	Brik with a pour screw cap	100% juice + A, C, E vitamins	Yes sport claim	Pear	No natural claim	Maxi pack
	Centre	PET with push and pull sport cap	100% juice + A, C, E vitamins	Yes sport claim	Pear	No healthy claim	No natural claim
	South	Brik with a pour screw cap	100% juice + A, C, E vitamins	Invigorating	No natural claim	No healthy claim	Maxi pack
Adults: male 17–45 years	North	100% juice	100% juice + A, C, E vitamins	Invigorating	Glass	Metal can	No natural claim
	Centre	100% juice	100% juice + A, C, E vitamins	Invigorating	Glass	Metal can	No natural claim
	South	100% juice	100% juice + A, C, E vitamins	Thirst-quenching	Maxi pack	Glass	No healthy claim
Adults: female 17–45 years	North	100% juice	100% juice + A, C, E vitamins	Invigorating	Glass	Metal can	No healthy claim
	Centre	100% juice	100% juice + A, C, E vitamins	Thirst-quenching	Maxi pack	Glass	No healthy claim
	South	100% juice + A, C, E vitamins	100% juice	Thirst-quenching	Metal can	Glass	No healthy claim

Practicing sportsmen: 15–25 years	North	100% juice + A, C, E vitamins	Juice and pulp + mineral salts	Invigorating	No sport claim	Glass	No natural claim
	Centre	100% juice + A, C, E vitamins	100% juice	Invigorating	No sport claim	Metal can	No natural claim
	South	100% juice + A, C, E vitamins	100% juice	Invigorating	No natural claim	No healthy claim	Glass
Practicing sportsmen: 26–40 years	North	100% juice + A, C, E vitamins	100% juice	Thirst-quenching	No natural claim	No healthy claim	Glass
	Centre	100% juice + A, C, E vitamins	Juice and pulp + mineral salts	Thirst-quenching	No sport claim	No natural claim	Metal can
	South	100% juice + A, C, E vitamins	Juice and pulp + mineral salts	Thirst-quenching	No natural claim	No healthy claim	Glass
Practicing sportsmen: 41–60 years	North	Invigorating	Yes health claim	100% juice + A, C, E vitamins	Nectar + ginseng	Apple	No natural claim
	Centre	Invigorating	Yes health claim	100% juice + A, C, E vitamins	Glass	Pear	No natural claim
	South	Invigorating	Juice and pulp + mineral salts	Juice and pulp + mineral salts	Apple	Apple	Glass

In this specific case, two very diversified segments with few, or no common points, can probably suggest the creation of two lines of drinks, each with a personal identity (companies, especially medium-large ones, often like this solution, while the smaller ones are more likely to find a compromise between the segments identified).

2.7 Why Search for Segmentation?

Individual differences in reaction to products and concepts are pervasive. During the period up to the 1940s and 1950s there were scarcities of items. Consumers were often happy to purchase what was available. Consequently, the Henry Ford belief that the consumer could have a car in any color, as long as it was black, was more common than believed. There was simply too much demand in contrast to too little supply. The notion that one might have to change the features of the product, or talk about the product in different ways to appeal to different audiences was simply not recognized as being valid. For every dissatisfied customer who did not like the characteristics of the product there might be three, four, or more individuals who would accept what was available. At this time the famous German college of design 'Bauhaus' under the director Hannes Meyer changed its focus from combining art and production to cost-cutting industrial mass-production to make products affordable for the mass of the population.

Recently for North America, and Western Europe, the assumptions of the mass market no longer hold true for most businesses and product categories. A firm adopts either a mass-market strategy or a market segmentation strategy. There is no in-between.[3] One variant of market differentiation is segmented marketing. The basic principle of segmented marketing is to allocate members in the consumer population as clearly as possible to at least two customer groups according to external criteria that can be specified. The bases for segmenting a market are nearly unlimited and can include such factors such as: product class behaviors, product preferences, brand selection behavior, demographics, geography, and socio-economic status. Indeed, the number of factors is almost limitless, although some factors generate reasonable segments whereas other factors do not.

Green and Tull (1978) set four basic criteria for market segmentation.[2]

(a) The segments must exist in the environment; *i.e.*, they are not to be a figment of the researcher's imagination.
(b) The segments must be identifiable (repeatedly and consistently).
(c) The segments must be reasonably stable over time.
(d) One must be able to efficiently reach segments through specifically targeted distribution and communication initiatives.

According to Neal, there are, broadly speaking, only two methods for segmenting a market: *a priori* and *post hoc* methods.[3] *A priori* segmentation is a procedure whereby a company chooses to break up customer groups by a generally accepted classification procedure assumed to relate to variations in customer purchase or use of the product category. This grouping may be the result of company tradition, recognized industrial groups, or some other external or internal criteria. Examples of *a priori* segments include such classification schemes as the following.

1. Standard Industrial Classification groups, (*e.g.*, www.govtsales.com/sics/sicgroups.htm).
2. Geographic regions or sales territories.
3. Basic demographic groups (*e.g.*, sex, age, household composition).
4. Purchase or usage groups (*e.g.*, heavy users, light users, non-users).

Despite the attractiveness of the segmentation criteria, demographic information cannot be used exclusively to determine customer needs. Consider an array of marketing articles that worked with this information about two individuals. Monahan asked: 'Are Grace Slick and Tricia Nixon Cox the same person?' Grace Slick, lead singer of Jefferson Airplane and Tricia Nixon Cox, the preppy daughter of former President Richard Nixon, were demographically indistinguishable. Both women were urban, working women, graduated from college, aged 25–35 years, at similar income levels, household of three, including one child. Demographics could not explain the distinctly non-demographic difference between Grace and Tricia.

More insight emerges about 'truer' segmentation through an analysis of an individual's 'mind-set'. One of these is the method of 'needs-segmentation: knowing *why* customer buys the product', can do more to help increase customer value than 'studying behavior: knowing what a customer buys'. The reason is quite simple. The motivation to purchase may be distinct for different customers who are buying the same product. One person buys it for him/herself and another one doesn't like the product but buys it for a friend as a gift. Neal rightly refers to an important caution that should be exercised when using the *a priori* criterion for segmenting a market. Our society is dynamic! This means segmentation studies conducted, or validated, 3 or 4 years ago may not be appropriate today.[3]

Post hoc segmentation is empirically derived on the results of research studies undertaken for the specific purpose of segmenting a market. Segments generated from such a study emerge after aggregating buyers who respond similarly to a set of 'basis questions'. Basis variables for such a *post hoc* segmentation study include examples such as product attribute preferences or product purchase patterns as well as benefits preferences or loyalty and socio-economic status, lifestyles or self-image.

Traditional algorithms for conducting traditional *post hoc* segmentation studies are:

1. cluster analysis
2. correspondence analysis
3. search procedures

4. Q-type factor analysis
5. AID-method (automatic interaction detector)
6. discriminant analysis
7. brand users profiles
8. assumption of a mass market.

Not all opinions are so positive about segmentation, however. Malcolm Wright from the Department of Marketing, Massey University refers to the misapplication of the expression 'market segmentation.' In his opinion, any set of competencies or capabilities will be adequate for market segmentation, because the term 'market segmentation' has come to be empty of any meaning.[4]

In the last 10 years there have been some interesting new developments in market segmentation research. The following sections discuss those recent advancements.

1. Multidimensional segmentation. This is a step-by-step process that begins by identifying customer segments from both strategic and tactical perspectives.
2. Artificial neural networks. Using sophisticated and specialized hardware and software, these networks attempt to simulate the multiple layers of simple processing elements called 'neurons'. These networks look for patterns that more conventional methods would not be able to discover.
3. Latent class models. These enable the user to simultaneously optimize a research function and find clusters of cases within that framework.

But if there are so many approved methods, why do market segmentation strategies sometimes fail? Neal suggests three reasons.[3]

1. A lack of senior management involvement and recognition that market segmentation is a strategy. A strategy must permeate the firm and the way it deals with the marketplace.
2. A lack of understanding of the concept of market segmentation and its need to identify groups that truly exhibit behavioral response differences to variations in the marketing mix.
3. A presumption that all markets can be segmented on bases that are subject to influence by variations in the marketing mix.

For the most part segmentation in the conventional methods do not reveal many differences in the ratings of concepts or products as assigned by the segments. That is, when the researcher divided the population into different groups of individuals, there appeared to be no clear differences between the subgroups. What appeals to one group appeals to the other group as well. Plotting the concept scores for one group against those of the other group shows a 45 degree line, across a large number of different concepts. Although people might differ from each other in ways that we consider dramatic, that difference does not manifest itself in strong differences in reactions to concepts.

Despite the recognition that segments are usually quite similar in the way they react to concepts, except perhaps in the most unusual cases where the concepts clearly are designed for one of the segments, researchers have kept hoping that the analysis of different subgroups would, in the end, generate different concepts. That hope continues today. Users of research continue to demand representation of different groups in the population, hoping and ever believing that they can find a relation between some conventional measure such as age, or brand used most often, and reaction to the concept. The relation does not appear. When the researcher works with full concepts, the lack of the relation is not troubling. It is similar to the lack of a relation between reactions to full products and brand used most often. The full product, like the full concept, is so complex that there is nothing to drive a strong expectation that a person who uses a specific brand would strongly react to the concept. There is so much going on in the concept that any relation between subgroup membership and response to stimulus is masked by the complexity of the stimulus.

The perplexing lack of relation between concept performance and segment membership developed by conventional means becomes quite obvious when the researcher systematically varies the concept in terms of different types of messages (*e.g.*, health messages *versus* taste messages). If a person says that he likes health products, we might expect the person to respond strongly to concepts that emphasize health. The person might or might not respond strongly to concepts that emphasize taste.

2.8 Applying the Segmentation Algorithm to Concepts or to Concept Elements

There is no reason, however, that the approach be limited to products. With some modifications the researcher can apply the approach to concepts as well. The segmentation approach, originally designed for products, requires that the evaluative attribute (*e.g.*, overall liking) be related to a descriptive characteristic of the product. This is the sensory attribute. It further requires that the same respondent evaluate several different products, usually at least six, in order to generate enough data points for a curve. The key to migrating the segmentation approach to concept work is to maintain a parallelism, treating concepts or concept elements as if they were products.

The parallelism defines the following set of equivalences.

1. Stimuli. Originally the stimuli in sensory preference segmentation constituted complete products. Now the stimuli are either complete concepts or concept elements from a conjoint study.
2. Dimensions. Originally the dimensions were sensory profiles of the products, obtained either from expert panelists in their descriptive

language, or from consumers using consumer terms. For concepts the dimensions are either communication profiles of complete concepts, obtained from consumers in the study, or semantic scale profiles of the individual elements obtained from a group of dimensionalizers. For both product and concept research the dimensions are not evaluative, but rather descriptive. Communication attributes can be scales such as 'more for men *versus* more for women', 'for adults *versus* for children', 'for taste *versus* for health', *etc.* These attributes may have a hidden hedonic attribute, but on the surface they appear to be simply descriptions of the concept or of the concept element.

3. Key evaluative attribute. For products the key attribute is overall liking. For complete concepts the key attribute is either overall liking or purchase intent, depending upon the specific scale used. For concept elements the key attribute is the part-worth utility for liking or for purchase intent, respectively. Each concept or each concept element has one value for this key attribute for the total panel.
4. Modeling. The modeling step remains the same. The objective of modeling is to relate the key evaluative attribute to the semantic or communication attribute, by means of a simple quadratic equation. For example, if the study deals with beverages, and the consumers or the dimensionalizers rate the concept or element on 'more for taste *versus* more for health' (abbreviated as TvH), the equations would be written (k_0, k_1 and k_2 are constants):

$$\text{Concept purchase intent} = k_0 + k_1(\text{TvH}) + k_2(\text{TvH})^2$$

$$\text{Conjoint element utility} = k_0 + k_1(\text{TvH}) + k_2(\text{TvH})^2$$

2.9 Establishing the Validity of Internet-based Research for Concepts

Sometimes the validity of a research method needs to be established by demonstrating that the same results will be obtained following two different methods. This definition of validity pervades the market research community, especially when the issue demanding validity consists of a new way of doing the same research, rather than doing a new type of research procedure. Validity with new field executions simply demands showing that the new method produces the same results, but either gives additional information, is easier to execute, or costs less money to execute.

References

1. G. T. Fechner, *Elemente der psychophysik*, Breitkopf und Hàrtel, Leipzig, 1860.
2. P. Green and D. Tull, *Research for Marketing Decisions*, Prentice Hall, Englewood Cliffs, NJ, 5th edn, 1978.

3. W. D. Neal, Principles of Market Segmentation, 2003, http://www.marketingpower.com.
4. M. Wright, RESPONSE - Competences/Capabilities for Market Segmentation, 2003, www.marketingpower.com.
5. J. L. Balintfy, W. J. Duffy and P. Sinha, Modeling food preferences over time, *Oper. Res.*, 1974, **22**(4), 711–727.

CHAPTER 3

Ideas from the World of Information Science: A Model-driven Development

PETRAQ PAPAJORGJI,*[a] ARDITA TODRI[b] AND ORKIDA ILOLLARI[c]

[a] Canadian Institute of Technology, Tirana, Albania; [b] University "Aleksander Xhuvani", Elbasan, Albania; [c] Mediterranean University of Albania, Tirana, Albania
*Email: petraq@gmail.com

3.1 Introduction

The history of computing is characterized by the increasing level of abstraction of the arsenal of tools needed to design and implement software systems. The higher level of abstraction used the easier writing code has become and complex software systems have been built faster and cheaper.[1]

One of the most important consequences of increasing the level of abstraction has been the net separation of the model, presenting concepts of the problem domain, from the code used to implement the software system. The separation of the model from the code made it possible for researchers and practitioners of software development to focus on finding tools and methodologies to design the model faster, better and able to generate the code in different programming languages or to directly debug and execute the models. Generating the code took care of all the plumbing that is necessary to execute the software, thus enormously facilitating the software construction process.[2]

Consumer-based New Product Development for the Food Industry
Edited by Sebastiano Porretta, Howard Moskowitz and Attila Gere

Published by the Royal Society of Chemistry, www.rsc.org

The modeling process has been largely improved by the advent of the Unified Modeling Language (UML), which nowadays is a standard for visualizing, specifying, constructing, and documenting all the steps of software design and development. UML is composed of a set of different diagrams that allows for presenting different views of the system under construction with a strong focus on static and dynamic aspects of the system. UML is used to design a platform-Independent model (PIM) that is a high level of abstraction model, independent of any programming environment. The PIM is created using concepts from the problem domain and their interrelations. The fact that it is computing platform independent makes PIM a lifelong intellectual investment that is insulated from technology changes.

At the final stage, the PIM will have to be translated into a system that will be hosted by a particular computing platform. Thus, different computing platforms may have different particularities that have to be taken into consideration while implementing the PIM. A PIM adapted for a particular computing platform is referred to as a platform-specific model (PSM). Therefore, the same PIM could be implemented in different computing platforms and result in several PSMs.[1] The structure PIM and potentially several corresponding PSMs creates a two layer system that keeps the business logic separated from the code implementation. Experience has shown that, in general, business logic has a much longer life than the implementation technologies. Technological changes should not impact the structure of the business model.

There have been several attempts to create environments that provide capabilities to create UML-based models that can automatically generate code and/or debug and execute models directly.[3,4] In this study the focus is more on the work of Lano[3,5,6] and his product UML-RSDS.[7]

UML-RSDS is a tool for model-driven development. UML-RSDS provides rapid automated construction of software systems (in Java, C, JSP/Servlets, C# or C++) from high-level UML specifications (class diagrams, use cases, state machines and activities). This tool can be used as part of software development, following the principles of model-driven development. The tool provides many facilities to the process of software development such as reducing coding costs, time and errors debug, and provides a greater consistency between system components written in multiple languages. UML-RSDS can ensure the correct and consistent implementation of business data and logic in multiple system components. This tool uses a simplified OCL (Object Constraint Language) notation for implementing constraints on modeling elements. In particular, two-valued logic is used, and all collections are either sets or sequences. This also facilitates verification in cases of high-integrity systems requiring a high degree of assurance.[7]

The UML model can be used to generate systems that can be defined using many different technologies and techniques. As a rule Java technologies are used, and three alternative system styles such as:

- Servlet style: presentation tier is coded using only servlets and auxiliary classes to generate web pages;

- JSP style: presentation tier is coded only using JSPs to handle web requests and to generate web pages;
- J2EE style: Java Entity beans and Session beans are constructed to define the business tier of an application, together with value objects and a database interface class. In each case, the system specification is given as a class diagram and set of use cases defining the user operations to be provided by the system. Use cases can be:[3]
  - (i) create E: create an instance of entity E (setting its attribute values)
  - (ii) delete E: delete an instance of E
  - (iii) edit E: edit an instance of entity E (setting its attribute values)
  - (iv) add E r: add an element to the role set r of an instance of E
  - (v) remove E r: remove an element from the role set r of an instance of E
  - (vi) list E: list all instances of E
  - (vii) searchBy E a: find instances of E with a given value for attribute a.

As a result, it can be said that developing a model has many advantages, such as separation of concerns, high level of abstraction, and converting the same model to different programming environments. Using a model-driven development approach as Lano suggests,[3] has more advantages as, based on a well-defined model, it can generate in a fraction of second, thousands of lines of tested and optimized code in several programming languages.

3.2 The Model

One of the most important tasks in the long list of tasks to be fulfilled while managing a software project is to have a deep knowledge of the problem to be solved. Thus, understanding the domain of the problem and depicting the concepts representing the domain is not an easy task. It requires long hours of dialog with the team representing the business part, and personal involvement of software engineers in understanding the problem domain. This understanding of the problem is expressed in a UML class diagram, in the form of a model representing the domain concepts, their attributes and behavior and the interrelations among concepts.[1]

The UML model shown in Figure 3.1, presents the concepts in the domain of Mind Genomics, their attributes and the relationships among them. The top element is the element *MindGenomicStudy* representing a study using the Mind Genomics approach. The study has a name that will be listed in the dashboard of site https://www.bimileap.com/u/dashboard. A study is conducted through an *Experiment,* which is another concept of the Mind Genomics domain. An *Experiment* has an attribute of type String referred to as *name*. The *Experiment* concept serves as a container of *Silos*, another concept in the Mind Genomics domain. A *Silo* is used to represent a pillar in the study; in other words, one of the main issues that the Mind Genomic study will address. As an example, in the study 'Distance Learning' undertaken using the Mind Genomics approach just after the Covid19 crisis

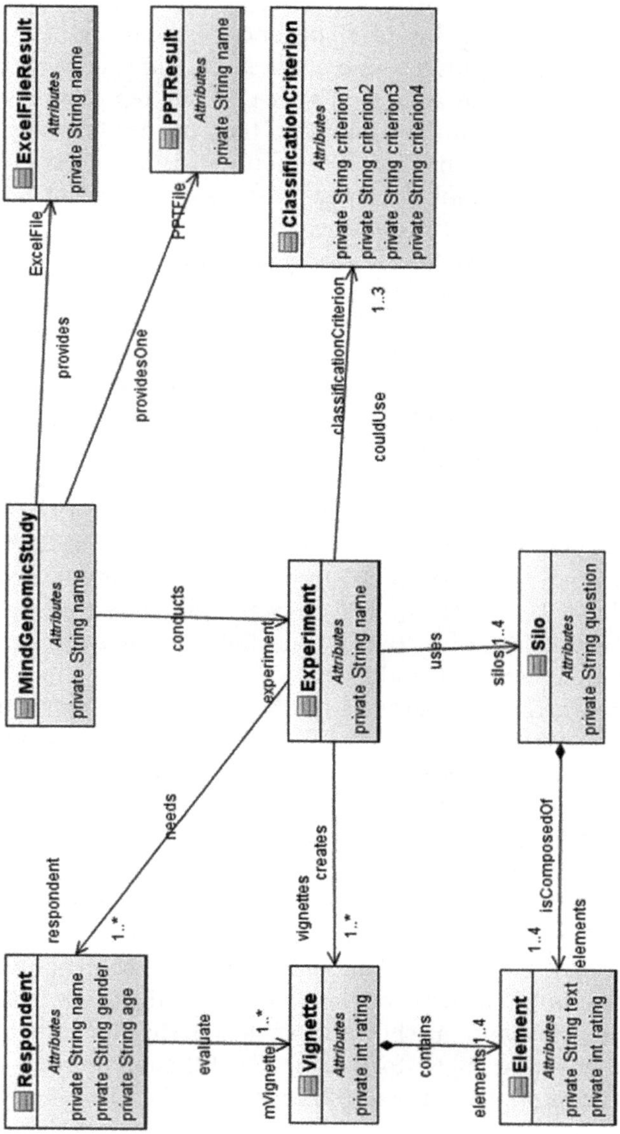

Figure 3.1 The model representing concepts and the relationships among them.

started and all academic institutions around the world were closed, feelings of students and professors were analyzed to evaluate how this new paradigm stands *vis-à-vis* the traditional approach to learning, the face-to-face approach. Effectiveness, interactivity, feasibility and the future of distance learning, respectively were selected as the pillars of the study.

The focus of this study were the following four pillars: effectiveness, interactivity, feasibility and the future of distance learning, respectively. Thus, the concept of the *Silo* represents one of the main points of the study, one of the issues the study should address. The issue a silo represents is presented in the form of a question. In the example of the Distance Learning approach, one of the silos is represented by Question 1 – How effective is distance learning? An experiment could have four to five silos. A navigable association referred to as 'uses' shows this constraint.

In the case of the Distance Learning study the silo representing the interactivity pillar was named as Question 2 – How interactive is distance learning? The silo representing the 'feasibility' pillar was referred to as Question 3 – How feasible is distance learning? The silo representing the future of the distance learning approach is referred to as Question 4 – What is the perspective of distance learning as a learning approach?

Another example of selecting the four silos can be found in Ilollari *et al.*,[8] the example presented relates to the feeling of clients regarding mobile banking in Albania. The first silo represented in a form of a question is Question A: What is the culture of mobile banking in Albania? The second silo is represented by Question B: How Albanian clients evaluate mobile banking services? The third one is Question C: What is the use of mobile banking in Albania? The last one is Question D: What are the pros and the cons of mobile banking in Albania?

A silo is composed of four answers for the question it represents. Answers to the question represented in a silo, are referred to as elements, instances of class *Element*. The *Element* is the basic unit of the rule developing experimentation known as RDE.[9] It is used to describe a simple idea that can stand by itself. It is not connected to something else in terms of grammar, but it is used as one of the potential answers of the question representing a particular silo. Table 3.1 shows the silos and the corresponding answers/elements for each silo for the Distance Learning study.

In the case of the study 'Understanding client's feelings about mobile banking in Albania' the answers or elements pertaining to the first silo are as shown in Table 3.2. Again, as shown in Table 3.2, the elements of a silo are simple statements, that tell a 'story' and each of them represents a potential point of view or response to the question/silo.

The relationship between concepts *Silo* and *Element* is *one-to-many*, meaning that one silo will contain four elements or answers. It is important to be precise about the nature of the relationship between two concepts as this information is considered when the code is generated based on the model. The same model could be used to generate code in several different programming languages. A UML model in the UML-RSDS system[3] could be used

Table 3.1 Example of silos and the corresponding elements/answers.

Question 1 – How effective is distance learning?	
A1	Distance learning is not as effective as face-to-face communication
A2	Distance learning is more an individual learning approach
A3	Distance learning platforms are less friendly to use than social media (Facebook, Instagram)
A4	Distance learning is more appropriate for people that have a job
Question 2 – How interactive is distance learning?	
B1	Distance learning platforms do not favor immediate feedback from students
B2	Distance learning platforms are not helpful for students during web-seminars
B3	Distance learning platforms allow for interaction between professors and students
B4	Distance learning platforms push students towards rational thinking
Question 3 – How feasible is distance learning?	
C1	Distance learning platforms require high-speed Internet
C2	Distance learning platforms operate with a limited supportive infrastructure
C3	Distance learning platforms do provide enough support for labs, seminars and exam sessions
C4	Distance learning platforms provide full support to forums, quizzes, messages, announcements, audio and video recordings
Question 4 – What is the perspective of distance learning as a learning approach?	
D1	Distance learning platforms require the ability to upload and download files on and off line
D2	Distance learning platforms should provide interaction as in the classroom
D3	An online tutor is necessary (besides the lecturer)
D4	A better distance learning approach requires more infrastructures and human resources

Table 3.2 Example of answers/elements to a silo.

Question A: What is the culture of mobile banking in Albania?	
A1	Clients are well familiar with functions of mobile banking
A2	Clients have good knowledge of banking services
A3	Clients are not familiar with banking services
A4	Clients are attentive to mobile banking advertisements

to generate code in C++, C#, C code, Python, and a few versions of Java such as Java 4, Java 6 and Java 7. *etc.* Thus, a *one-to-many* relationship means that in the starting class an attribute of the type *ArrayList* contains objects of the destination class. As an example, in the body of class *Silo* an attribute of type *ArrayList* containing instances of class *Element* will be declared.

The association linking two different classes has a particular meaning. In the case of the association *isComposedOf* between classes *Silo* and *Element* shows the fact that a silo is composed of elements. In Figure 3.1, this association is navigable towards class *Element*; meaning that from class *Silo* one can reach data and behavior in class *Element*. This association being of type *composition* means that a silo (the whole) is composed of elements (parts); it is an association of type *whole-part*, showing that the *whole* is composed of *parts*. A composition has a strong sense of ownership between *parts* and the *whole*; a *part* can belong to only one *whole*.

Figure 3.2 Example of a vignette.

Another important concept in the Mind Genomic domain is the one of *Vignette*. A vignette is created by a combination of elements according to the RDE rule.[9,10] Thus, a vignette could be composed of 2–4 elements, as shown in the model of Figure 3.1. A vignette has one attribute referred to as *rating* of type Integer that takes values from a Likert scale with a 1–9 selection. A vignette, as the diagram shows, is composed of up to four elements following a strategy designed by RDE. Figure 3.2 show an example of a vignette for a study on Recycling Luxury Items & Accessories. As shown in Figure 3.2, the vignette is created using elements or answers from different silos and presented for evaluation to a respondent. Figure 3.2 also shows the Likert scale presented to the respondent for vignette evaluation.

Vignettes are presented for evaluation to a *Respondent*. Class *Respondent* has three attributes such as name, gender and age that are used as criteria to group respondents. These elements allow the system to study differences in judging the problem between different gender and group ages. Different respondents will evaluate different vignettes and a respondent will never evaluate the same vignette twice. A respondent will evaluate 24 vignettes in an experiment with four silos and four elements per silo.

While designing the *Experiment* the system offers to include a three or four Classification Criterion that is used to classify respondents into different groups. Therefore, the system has the class *ClassificationCriterion* class to represent this concept. This class has four attributes of type String to hold the values for each criterion.

After collecting the data from the respondents, the system uses complex statistical models, data mining and clustering techniques to analyze the data. The results of data analyses are returned to the researcher in the form of two data files: the first file in Excel format has all the results distributed into different tabs and the second one in the PPT format. These results are shown in the model with classes *PPTResult* and *ExcelFileResult* respectively.

The Power Point File generated automatically by the system is a 45-slides presentation about the topic in the study. A few slides describe the Bimileap system, its approach, how Bimileap tries to contribute to society, offering its services free of charge, *etc.* Later the focus goes on to describe the current study, what researchers would like to achieve and the means of reaching the required results. The rest of the slides present the results of the study and insights into how to interpret these results.

Another Excel file is provided by the Bimileap system with detailed results combined by age, gender and the preliminary questions selected at the beginning of the study. Any Excel tool could be easily used to display the obtained results through different types of graphs.

3.3 Conclusion

Model-driven development is an approach that considers models as the first-hand elements for designing and implementing software systems. It starts with the design of a model that presents all concepts of the problem domain, their attributes and their interrelations using associations. A model stands above all computing platforms and thus has to be presented using a high level of abstraction. An abstract model has a longer life and is resilient to changes in technology. A model could be translated into different computing platforms.

Presenting the concepts of the domain in a model makes it easier for developers to implement it in a particular programming language. A model is the starting point for any future developments or improvements as the model is general and independent of any computing platform. Reading and presenting the model to future stakeholders does not require any previous knowledge of programming languages or software engineering experience. The model bridges software engineers and business people, thus making their communication easier.

References

1. P. J. Papajorgji and P. M. Pardalos, *Software Engineering Techniques Applied to Agricultural Systems An Object-Oriented and UML Approach*, Springer, 2nd edn, 2014.
2. L. Starr, A. Mangogna and S. Mellor, *Models to Code: With No Mysterious Gaps*, APRES, ISBN-13: 978-1484222164, 2017.
3. K. Lano, *UML-RSDS: A Tool to Support Agile Model-based Development Using UML*, CRC Press, London, 2017.
4. L. Starr, *Executable UML and Action Language blog*, Mars, 2015. https://modeling-languages.com/leon-starrs-executable-uml-and-action-language-blog/.
5. K. Lano and S. Y. Tehrani, Experiences of teaching model-based development, *IEEE 18th International Conference on Model Driven Engineering Languages and Systems_Educators Symposium (EduSymp)*, Ottawa, Canada, 2015.

6. K. Lano and H. Haughton, *Financial Software Engineering. Topics in Computer Science*, Springer, Chester, UK, 2019.
7. K. Lano. *The UML-RSDS Manual*, 2018, https://nms.kcl.ac.uk/kevin.lano/uml2web/.
8. O. Ilollari, P. Papajorgji and A. Civici, Understanding client's feelings about mobile banking in 2nd UNICART, *Interdisciplinary International Conference on Management, Tourism And Development of Territory*, Albania, 6–7 April 2020.
9. A. Gofman, *Rule Developing Experimentation in Consumer-Driven Package Design*, https://doi.org/10.1002/9780470400531.eorms1036, 2011.
10. A. Gofman and H. Moskowitz, *J. Sens. Stud.*, 2010, **25**(1), 127.

Part Two: Emotions and Consumer Testing with Children

CHAPTER 4

Imputing Emotions to Foods

ARIOLA HARIZI

Slovak University of Agriculture, Nitra, Slovakia
Email: ariolaharizi@hotmail.com

4.1 Introduction

Many studies have been conducted in the last years demonstrating how consumer behavior intertwines with marketing to drive sales. These factors vary in their nature; the situation, personal and psychological factors *versus* the economic factors, and so forth. The range of variables and the nature of the interactions become further complicated by the varieties of control that can be exerted by the manufacturer. These factors range from the product itself to the control implicitly placed upon the family. For example, the former, product, comprises, product quality, *e.g.*, product pricing, and product advertising. The latter, implicit control of the family, focuses on what the family can afford, the life stages of the family members, and the media consumed by the different family members.

Beyond the manufacturer and the family lies the world of the person, the psychology of the consumer as an individual. There is a growing interest in the link between the personality of a person and the person's purchase and consumption behaviors with respect to food. Food is especially interesting because the need for food purchases recurs again and again, over a short period of time. Furthermore, food occupies a lower level in the person's hierarchy of needs. According to psychologist Abraham Maslow, the need for food is lower than the need for cars and other material and non-material 'things' and 'experiences'. At the same time, although the need for food is lower on Maslow's scale, there may be a higher-level aspect, *i.e.*, the nature of the sensory experience that can be imparted by food.[1]

Consumer-based New Product Development for the Food Industry
Edited by Sebastiano Porretta, Howard Moskowitz and Attila Gere

Published by the Royal Society of Chemistry, www.rsc.org

Human senses, consumer experiences and sensations are considered in emerging marketing paradigms as a major topic for study. They provide a complex of individual topics rich with potential for the world of consumer research on the one hand and food research on the other. Sensory experience based on consumer behavior has been exploited by many marketers to add both a pleasant experience with food as well as to create an emotional link between the person and the food. There is an emerging field of knowledge about sensory links to experience that will be summarized in this review. The main objective is to review the effects of sensory marketing on buying food products, and to introduce a new paradigm of 'imputed emotions' to the experiences involved in the purchase process, *i.e.*, 'buying'.

4.2 The World of Today

In today's world abounding with new foods, yet with many foods similar to each other, marketers are recognizing that a new direction must be carved out, beyond simply a modified, hopefully attractive and engaging description of product features. Today, the notion of 'emotion' in behavior is emerging, coming to the forefront of business, and thus becoming a new topic of research. Emotion links closely to experience. In the world of food-related studies involving sensory and marketing analysis, the focus must include the experience of consumption, but equally important and yet unexplored, the experience of preparing for consumption.

The many published studies in the world of food on the one hand, and consumer science on the other, suggest the importance of psychological factors in the selection of food at the store, as well as the venues where consumption occurs, *e.g.*, in the home, in restaurants, as well as on-the-go snacks, at work, and in school. If the scientific literature is not enough to prove that, a search through Google reveals 352 million hits for the search phrase 'factors driving selection of food'. When we move to Google Scholar® we drop down to a mere 973 000, almost a million hits.

People appear to be rational in their selection of food, or it at least that is what many marketers believe, as do many adults. Adults are presumed to know what they are doing, and to be aware of the consequences of their actions. This experience of controlling one's own actions and, through them, the course of events in the outside world is called 'sense of agency'. Sense of agency forms a central feature of human experience, although surprisingly it is only recently that the brain mechanisms responsible for this important aspect of adulthood have begun to be investigated systematically.[2] This sense of agency is a central aspect of human self-consciousness and refers to the experience of oneself as the agent of one's own actions. As yet, however, there is only a tenuous link between the psychobiology of agency and the day to day experiences of people.

Sensory marketing is a marketing strategy that focuses on the senses to influence the consumer's feelings and behavior.[3,4] Traditional marketing views consumers as rational decision-makers who care about functional features and benefits. In contrast, however, experiential marketers view consumers as rational and emotional human beings who are concerned with achieving pleasurable experiences.[5] All we need do in order to understand the power of sensory marketing it to look at the effect of packaging, the first experience a consumer has with the product, even before the product is prepared and eaten. Packaging is a critical aspect of the marketing offer, with many implications for the multi-sensory customer experience. Packaging carries messages that drive experience and feeling. Packaging affects attention, comprehension of value, perception of product functionality, and also consumption, all with varying consequences for consumer experience and response.[6] Understanding the 'What' consumers want as well as the 'Why' are two of the most significant hurdles faced by any business creating products for consumers. Sensory research helps with the effort to identify factors and issues that contribute to a product's success or failure.[7]

Experience is becoming recognized as a tactic to drive sales. Marketers are increasingly allowing consumers to sample sensory-rich experiential products before purchase.[8] A customer differentiates the product by sensory characteristics, and quite often the customer is attracted by the brand because of the sensory characteristics of the brand, and thus the sensory experience enjoyed when consuming the product.[9] Sensory branding in this context can actually move from product to country, so that one can attach the product to the country, and while 'selling' the product, one sells the country connected to the product.[10] In experimental psychology this is known as 'secondary reinforcement'. The product is the primary reinforcement because it is enjoyed on a sensory basis. The country paired with the product becomes reinforcing as well, because of the pairing.

For common foods, sensory characteristics alone do not suffice to drive interest, purchase, and consumption. Many other non-sensory factors come into play. The matrix of information and experience is not limited to the sensory characteristics, nor even to the physiological effects, but generally to a far greater array of factors, such as previous information acquired about the product, past experience, and attitudes and beliefs.[11]

4.2.1 Sensory Marketing Using Five Senses

Customers want products 'that dazzle their senses, touch their hearts and stimulate their minds'. Practitioners of sensory marketing distinguish five types of experiences, labelling them 'strategic experiential modules (SEMs)'. These experiences make sense to us as consumers and as scientists. The experiences cover a variety of impressions. The five experiences are: sensory experiences (SENSE); affective experiences (FEEL); creative cognitive experiences (THINK); physical experiences, behaviors and lifestyles (ACT); and social-identity experiences which emerge by relating to a reference group or culture (RELATE). These experiences are implemented through so-called experience providers (ExPros) such as communications, visual and verbal identity, product presence, electronic media, *etc.*[12]

The five senses enable us to receive information from the environment in the form of sound, light, smell, taste, and touch.[13] Using the five senses at stores to showcase the product and drive sales illustrates an important opportunity for a 'marriage' between brand and consumer, and the chance for the brand to differentiate itself from competitors.[14] The strategy of sensory branding is to use all the five senses, simultaneously when possible, and by so doing, create a unique, engaging, and effective five-dimensional experience for the consumer.[15] Add to the five-dimensional experience the creation of pleasure, of product acceptance, 'liking', and one has an opportunity to take advantage of market opportunities. It is within this framework of creating the appropriate, effective sensory experience that food scientists find their greatest opportunity. Consumers are the final step in the production chain; it is useful to identify which factors affect their behavioral patterns.[16] Food scientists, with their expertise in assessing and using the sensory characteristics of foods, make an important contribution to the ability of food producers to predict consumer preference and choice.[17]

Worthy of note in the discussion of sensory marketing is the world of children and youth, the younger generation. Food marketers are interested in youth as consumers because of their spending power, their purchasing influence, and their role as future adult consumers. Multiple techniques and channels are used to reach youth, beginning when they are toddlers, to foster brand-building and influence food product purchase behavior. These food marketing channels include television advertising, in-school marketing, and product placements for kids products.

Children's food choices are influenced by the media, children of all ages watch television and everyone wants to eat all the fast food shown on the television. Disney, Nickelodeon, and PBS channels advertise different brands of food, such as McDonald's, Wendy's, Chuck E. Cheese; all of these brands sell fast food. These brands attract children of all ages. These companies sell their products to children in many ways. For example, children are encouraged to buy Happy Meals to collect toys. This marketing strategy has attracted research interest among those interested in the welfare of children. For example, one study involved pre-test and post-test measures on 1522 eligible students. The results suggest that cumulative exposure to TV food advertising promotes beliefs and attitudes supportive of those foods most heavily represented in food advertising on children's TV. The study links the consumption of fast foods and sweet drinks presented on children's TV with the consumption of such foods.

4.2.2 The Five Senses and Beyond

Today's food and beverage industries face an intensely competitive business environment. To the degree that the product developer and marketer, as well as the general business manager, can more fully understand the consumer and target development and marketing efforts, it is likely that the business will be more successful, everything else held constant. Among the first business-oriented books for today's hyper-competitive world we find '*Sensory and Consumer Research in Food Product Design and Development*', a multi-authored book which presents from the business viewpoint the critical issues faced by sensory analysts, product developers, and market researchers in the food and beverage arena.[18]

In psychology, the dominant computer metaphor, 'information processing', has been challenged by researchers demonstrating various manners in which mental activity is grounded in sensory experience and is not simply in a world of its own.[4] We now deal with each 'folder' of human senses to see how the human senses have driven the marketing effort through sensory experience. The topic is called by the newly coined phrase 'brand sense'. Brands can create emotional associations in the customer's mind by appealing to their senses. Sensory branding is a type of marketing that involves and appeals to all the senses (sight, smell, touch, sound, taste) in relation to the brand. At the level of practice, sensory branding has evolved to be one of the most important marketing tools used by the most successful companies like Apple, Starbucks, Singapore airlines, *etc.*[19]

Our focus will be on sensory branding beyond just the food itself. The deeper understanding of the food in terms of its sensory properties can direct marketing decisions concerning not only products, *e.g.*, product positioning with respect to competitors, but also market segmentation, customer relationship management, advertising strategies, and price policies.[20]

It is important to note here the sparseness of scientific literature regarding sensory marketing. The sparseness can be explained by the fact that most of the sensory marketing efforts are tactical, based upon momentary and often simply creative inspiration. Sensory marketing itself does not invite detailed experimentation as much as one-off events. On the other hand, we can learn a great deal from the nature of these events, simply because they have been thought through by experts, considering the nature of event feasibility and event success.

Up to now, most methods in the world of sensory analysis have focused on the study of the product itself, as a physical entity, to give shape to a product and to ensure its perceived product identity and quality. There is an expanded role for the world of sensory analysis, however, one which will need a technical contribution. This world will need to know the science of sensory impressions as it moves towards creating new foods and new food-experiences through the selection of new materials. This is the emerging area called 'new integrated product materials selection (IPMS)'. The models for IPMS incorporate factors such as fashion, market trends, cultural aspects, aesthetics, and recycling, as well as the target group.[21] It is not that these factors were never considered by sensory analysts, rather that the relations between the sensory properties and these factors were simply never thought of as being worthy of formalization through deep study.

A few examples will show how sensory marketing is executed. The first example comes from the marketing efforts of the well-known chain, Dunkin Donuts, now simply Dunkin. In South Korea, Dunkin Donuts used sensory marketing by releasing a coffee odor into the air, producing a pleasing coffee smell, all the while playing a company jingle. There are less bold tactics of this type for those who feel that the Dunkin Donuts strategy was a bit bold and perhaps extreme. There are other ways to play on the customers' sense of smell. Two alternative tactics are to place scented advertisements in major magazines or use a trademark fragrance, letting the fragrance waft in the store as the shoppers go about their business, inspecting, selecting, and buying. Studies have shown pleasant smells can improve mood (sales? or mood? by 40%. The lesson is to 'find a way to incorporate scent into your brand and stick with it'.

Sound is also important in sensory marketing, allowing the brand and the associated experience to create emotional appeal, and thus bond with the customers. Sound takes many forms in advertising. Brands like Airbnb do this beautifully, with their 'Belong Anywhere' campaign. It can be State Farm's 'Like a good neighbor...' jingle or a slogan that one can simply not 'get out of one's head', in the popular description of the power of sound. Consider lingerie giant Victoria's Secret, for example, which plays classical music in its stores to make customers feel part of an exclusive shopping experience. Today, sensory advertising requires that all brands find a way to use sound in their branding efforts, whether the effort is directed toward advertising (Airbnb, State Farm) or the actual shopping experience (Victoria's Secret).

In an effort to surprise and delight tennis fans at this year's Australian Open, the grocery chain Woolworths is hosting an exclusive event called the 'Woolworths Summer Sensorium'. Instead of serving up something that is simply tasty, the chain has developed a way to engage all five senses at once, in order to demonstrate how much of a difference freshly prepared food from a brand such as Woolworths can make. The Woolworths Summer Sensorium is an experiential event taking place during the Australian Open Tennis Championships. Between 16 and 29 January, fans will be able to sit down for a dining event called 'A Taste of Australian Summer', which is hosted inside a giant geodesic dome. During the event, attendees will enjoy freshly pressed juices and three paddock-to-plate, summer dishes taken from Woolworths' FRESH magazine. While eating, attendees will enjoy immersive 360 degree projections narrating the journey of each ingredient used to prepare their meal. By pairing immersive visuals with freshly prepared food, the event serves as a unique way to showcase Woolworths' amazing products and some of the delicious recipes in its FRESH Magazine.

Sight

Sight is, of course, one of the most powerful senses in the world of advertising. Luckily, it is also easy to work with. To play up the value of sight, one might start with the corporate website, making sure that it is attractive, engaging customers. Professionals design websites, as well as designing the social media and advertising graphics. At the more basic level, dealing with the visual impression itself, the specific design features, especially color, can be associated with the product or service. Giant brands such as Coca-Cola and Virgin Airlines have found ways to use color to promote brand recognition for their products. Sight is the most developed of the senses as 83% of the information that we receive is *via* sight. It not only plays an important role in key retail moments, such as whether or not the customer enters a shop, but it also responsible for transmitting the brand image and values. The customer's in-store visual experience is a combination of different stimuli including the store facade, the window display as well as the store's interior where the design, furnishing, lighting, and color choices all come together.

With a graphic simulation of the ocean, the 'Hugo Boss Sailing' video By Lars Kubric is impressive. Commissioned by Schokolade Filmproduktion the visuals are to be featured at the new Hugo Boss flagship store in London. Directed by Florian Rederer, in collaboration with art director Andreas N. Fischer, their sailing montage takes on fluid simulation. With background sounds by musician Clams Casino, this rendition is hypnotic in its multi-sensory appeal. The oceanic motif is complemented by a striking metallic palette that gives their work a cool modernist appeal. With their stunning usage of the Blender3d program, the spherical otherworldly protagonist appears to drip the silver matter with an eerie sense of realism. These mind-bending visuals are modernist in their sleek aesthetic, cool-toned palette, synthesizer sounds and utilization of graphic technology. This is bound to lure a neoteric consumer to the Hugo Boss brand.

A truly modern marketing campaign, Mercedes-Benz #Untamed – A Digital Photo Installation, incorporated multi-sensory art, crowdsourcing through social media and distinctive branding that made for a fascinating concept. The German car company invited Instagram users to submit their most strikingly original photographs, generating interest in order to stage the #Untamed installation. Combining the crowdsourced photography with the ideal of transcending norms, Mercedes-Benz created a multi-sensory exhibit in which the Instagram images took over an entire room, submerging visitors entirely in each photo. By allowing visitors to experience this extraordinary installation, the company has created a branding message that showcases the new Mercedes-Benz CLA automotive model's 'unrivalled dynamism and untamed character'. Mercedes-Benz #Untamed – A Digital Photo Installation, epitomizes dynamic marketing that engages consumers through artistic originality and social media.

In order to inspire calm and educate the public on life under the sea, Tetra and the Aquatic Design Centre teamed up to create a pop-up aquarium in the London Underground. 'The Tetra Tranquility Tank' is an installation that gives commuters the chance to look at a collection of crabs, catfish, turtles, and angel fish, as well as a variety of natural plant life. The tranquil experience of looking at these beautiful aquatic creatures is complemented by a multi-sensory experience filled with therapeutic underwater sounds, calming lighting, and the scent of freshly brewed herbal tea. In all, the experience is designed to help the busy people of London pass through the station and leave feeling more refreshed than when they first entered.

Hearing

Hearing is one of the major senses and like vision is important for distant warning and communication. It can be used to alert, to communicate pleasure and fear. It is a conscious appreciation of vibration perceived as sound.[22] Most verbal communication occurs in contexts where the listener can see the speaker as well as hear him.[23] Qualities of hearing experience include ease of listening, and the naturalness, clarity and identifiability of different speakers, different musical pieces and instruments, and different everyday sounds.[24] According to one study, the tempo of songs in a supermarket can affect a customer's buying behavior. Slow music in an experiment caused revenues to increase by 38% in a supermarket, and 41% in a restaurant. Faster songs also led to changes in customer behavior, leading to higher levels of stimulation, and encouraging visitors to act more friendly towards members of staff. Of course, it's not just music that speaks to your audience.[25]

Another two examples come from Starbucks and Dunkin (originally Dunkin Donuts). Both of these examples show the power of both experience (Starbucks) and advertising (Dunkin). The songs played in Starbucks stores are sent directly to each location by the company's main office, so everything remains a cohesive part of the complete brand identity. Even the artwork and wallpaper in Starbucks is the same wherever one goes. This way, Starbucks customers don't just get a great cup of coffee, but rather enjoy the full Starbucks experience at every store. Starbucks kept that concept in mind when they invested in a new marketing campaign with Spotify, designed to give customers the Starbucks 'audio' experience at home.

There are a number of multi-sensory experiences that have been created as of late that are tapping into touch, smell, taste, and sound in order to enhance the enjoyment of a product. On this note, Stella Artois recently collaborated with The Roots for the creation of two songs that were specifically created to enhance the sweet and bitter notes of its beer. Stella Artois describes its creation as 'the first music video you can taste', since it is formed from sounds that are associated with the flavors that can be found in Stella Artois beer. The two versions include one with higher pitches to bring out sweet notes, as well as one with lower pitches to specifically draw out the bitterness of the beer. In the past, Stella Artois has also experimented with other experiences where the senses collide, such as its 'Sensorium' dining event.

San Francisco restaurants are introducing curated playlists compiled to match their varied menus. The Ne Timeas Restaurant Group owns three restaurants in the Bay Area: Flour + Water, Central Kitchen, and Salumeria. They give out musical programs so patrons know what they are listening to while they eat, in addition to the standard food menu, drink list and dessert offerings. The atmosphere-creating selection comes from the Ne Timeas music directors. Each restaurant has a different vibe, which the curated playlists reflect. Though other eating establishments also carefully choose the music they play, Ne Timeas plays full albums and informs their diners the order in which they hear them. Earlier customers could be listening to something quite different than those who come in later in the night, so it is an important distinction.

Touching

Touch is the ability to perceive the world through physical contact.[26] The somatosensory system decodes a wide range of tactile stimuli and thus endows us with a remarkable capacity for object recognition, texture discrimination, sensory-motor feedback, and social exchange.[27] The sense of touch seems to carry information at the same time about the external object touching our skin, and also about our body itself.[28]

Prior research has assumed that touch has a persuasive effect only if it provides attribute or structural information about a product. Under this view, the role of touch as a persuasive tool is limited. People who are motivated to touch do so because the touching activity is fun, interesting or both. In turn, communications that incorporate touch lead to increased affective response and increased persuasion, particularly when the touch provides neutral or positive sensory feedback. People who are not motivated to touch for fun may also be persuaded by a communication that incorporates touch when they are able to make sense of how the touch is related to the message.[29]

A good business example comes from Apple. Apple as a brand is deeply focused on the digital and technological world. Despite this, they refuse to sell their products exclusively online. Instead, the business offers customers a way to interact with them through highly immersive in-store experiences. The stores, decorated in Apple's signature white color, are a visual treat, packed full of devices that one can touch, play with, and experience. The visiting customer can talk to Apple 'Genius' employees and listen to their opinion and enjoy an in-person interaction that is just as clean and fresh as a moment spent on the Apple website. Apple's shops are a crucial part of their sensory marketing strategy, and a way to introduce people to the lifestyle that Apple devices advertise.

Touch provides us with an important amount of information to make an informed decision. It is a sense capable of generating an experience that is more connected and interactive for the customer. The almost 500 Apple stores worldwide are an example of a store that drives touch marketing by allowing customers to use, feel, and try its products when in-store. This enables the customers to evaluate the quality and functionality of the Apple products at the same time, the experience of which drives both purchase and strong brand loyalty.

Pernod Ricard is now launching its first-ever interactive liquor shop retail space aboard the Cunard Queen Mary II cruise liner so as to deliver more personalized and educational shopping experiences to consumers. The store will feature a range of premium spirits from across the Pernod Ricard brand, including Royal Salute, Chivas, and Martell. Walls in the retail space are divided up into 'bays', each of which offers a different theme, like Cocktail Time or Celebration, that simplify the process of product selection. Interactive elements come into play with interactive screens that deliver detailed information on products, promotions, and competitions from each of the represented brands. Consumers will also be able to sample and learn more about the products available from Pernod Ricard with the in-store liquor tasting bar aboard the cruise liner.

Smell

When eating or drinking, a person experiences a multitude of sensations, including taste, smell, touch, temperature, sight, sound, and sometimes pain/irritation. This multi-faceted sensory experience is the underpinning of perceived flavor, although certainly some sensations contribute more than others.[30] Of all these, the sense of smell is probably the most important to drive the perception of what we call 'flavor'. Olfaction, the sense of smell, plays a far larger role in the appreciation of food flavor than is commonly thought. Smell is experienced through the nose, but also through the passage of air to the nose from the back of the mouth. At the same time, the sense of taste operates as does the sense of touch, taste coming from the stimulation of taste buds on the tongue, touch, then pain, temperature, and so forth from the stimulation of other nerves, in the mouth, such as the trigeminal nerve. These nerves contribute to taste and smell in a variety of food flavors such as the tingle of carbon dioxide in soft drinks, the bum from chili pepper, the bite of horseradish and so on. A common misbelief arises in that many volatile (gaseous) flavor materials in the mouth are actually smelled, but this is usually misinterpreted by untrained tasters as a gustatory sensation.[31]

The consumer product industry (perfumes, food and beverage, and pesticides) devotes millions of dollars each year to supporting consumer-oriented olfactory research in an effort to enhance product design and marketing. We are familiar with the importance of the evaluation of odors for applied purposes, but there is another area uncovered by sensory marketing. This area is anthropology, focusing on the role of odor in behavior. With so many intersecting areas of research, those who focus on the intersection of olfaction and anthropology have a tremendous contribution to make to this growing body of work that crosses traditional disciplinary lines. Anthropology could benefit from studying the powerful role of the olfactory system in memory, behavioral and social cues, evolutionary history, mate choice, food decisions, and overall health.[32]

The sense of smell mainly plays a priming role in eating behavior. It has been demonstrated that (orthonasal) odor exposure induces appetite specifically for the cued food. However, the influence of odors on food choice and intake is less clear, and may also depend on awareness or intensity of the odors, or personality traits of the participants.[33] Companies that use tactics from sensory marketing have discovered that the smell of cinnamon is often associated with feelings of warmth. In one study, adding a cinnamon scent to a cushion helped to improve the perceived effectiveness of its heat-based therapeutic attributes.[41] It is these discoveries, often by accident or by received wisdom from practice, that make up the knowledge repository of today's sensory marketing.

The sense of smell is stimulated unconsciously by fragrances and affects a person's mood either in the positive or negative direction, indicating scents ability to contribute to a sensory experience. Research regarding scents, as sensory stimuli, shows that the introduction of different scents with regard to both the brand and the service environment affect both sales and profitability.[34]

Care hire service Holiday Autos recently set up a fun vending machine that would dispense the smells of summer in place of ordinary snacks. The Smells of Summer vending machine features bottles that contain the smells of things like freshly cut grass, barbecues and the beach. In order to get a whiff of these intoxicating summer scents, people passing by were required to send out a tweet in order to redeem a bottle. The interactive campaign really speaks to a move brands are making to foster long-lasting connections with consumers through multi-sensory engagement. While neither scent-filled bottles nor social media vending machines are new ideas, Holiday Autos demonstrates the success that can come from converging two popular ideas for the purposes of sales and marketing.

Selfridges' Fragrance Lab takes shoppers through a multi-sensory journey, aiming to help them discover a personal scent and, in the process, help Selfridges better understand its consumers' behaviors, characters, and motivations. The experience begins with a personality quiz on an iPad, asking for information like a customer's name, age, shopping habits, and interests. From there, the experience guides shoppers through a series of rooms in the Fragrance Lab with audio, ultimately leading to a location where people can speak to a Selfridges' team member to further whittle down some options and try on some fragrances suggested just for them. The Fragrance Lab is part of Selfridges' Beauty Project, which started on 1 May 2014 as a 6-week campaign that will include a variety of events, discussions, and experiments like this one on the topic of beauty.

Dunkin is one of the best sensory marketing examples to discuss for insight into how engaging the senses can have an impact on your bottom line. Like Apple and Starbucks, Dunkin frequently infuses sensory marketing efforts into their brand, with carefully chosen packaging, specific brand colors and more. The brand even has its own jingle in certain parts of the world. In one sensory marketing strategy created within Seoul, South Korea, the company decided to make the most of their ability to encourage purchases through the sense of smell. Commuter buses around the city were given machines that emitted the scent of fresh coffee when the Dunkin song played on the radio. The results were impressive, leading to a 29% increase in brand sales.

Tasting

Information about a product may shape consumers' taste experience.[35] Taste, has a clear role as a (macro) nutrient sensing system, during consumption.[36] Conceptual framings of alternative food systems have been based principally on the reflections of academics and policy specialists rather than on the views of the producers and eaters who constitute the bulk of the food localization movement.[37] Initiatives are now emerging intended to create alternatives to the conventional, industrialized, global food system.

Heightened preference for sweet-tasting foods and beverages during childhood is universal and evident among infants and children around the world. The liking for sweet tastes during development may have ensured the acceptance of sweet-tasting foods, such as mother's milk and fruits. Moreover, recent research suggests that liking for sweets may be further promoted by the pain-reducing properties of sugars.[37] Five different sources of food emotions are proposed to represent the various reported eliciting conditions: sensory attributes, experienced consequences, anticipated consequences, personal or cultural meanings, and actions of associated agents.[38]

Remarkably, Ikea is an example of a retail brand that has achieved a strong taste association for its brand *via* its in-store restaurants, with 30% of Ikea customers now visiting the store just to eat. In addition to extending the customer's store visit, Ikea's restaurants also increase the purchase not only of food but also of other Ikea products. In fact, due to its popularity among consumers, a standalone Ikea restaurant with its famous meatballs could be coming to a neighborhood near you, the reader, in the not so distant future!

Starbucks is another pioneer among companies which uses sensory marketing. Over the years, the Starbucks brand has evolved significantly. The organization has become more than just a place to drink coffee – it has become a hub of creativity, intended to satisfy and stimulate sense of taste, sight, touch, smell, and sound. From the moment the customer walks through the door of a Starbucks venue, the customer can expect to smell the aroma of coffee, choose from the same iconic flavors, and hear a consistent style of music playing in the background.

When one thinks of enjoying a beer, the sense of taste is what often comes to mind first – but Norrlands Guld Ljus is now proudly stating that it has invented an entirely new way for people to 'drink' beer through their ears. The 'Ear Beer' was created to inspire ASMR, or an Autonomous Sensory Meridian Response, which is described as a euphoric experience that happens in the body when pleasing sounds are heard. Norrlands Guld Ljus' unique Ear Beer recording was shared to YouTube, Spotify, and Apple Music, as well as released as a limited-edition set of vinyls. As well as appealing to those who have an appreciation for sounds that trigger ASMR, the Ear Beer was paired with music from popular DJs and launched Sweden's largest music festival, Way Out West.

One surefire way to tell if a beverage has alcoholic content is by sniffing it, but the Ritz-Carlton Berlin is looking to change that by crafting perfume-inspired cocktails. Beginning on 12 June, Fragrances will add a multi-sensory mixology experience where the luxury hotel's guests can choose their drink by scent. As smell and taste are biologically intertwined, this German experiment aims to redefine taste. Conceived by Arnd 'The Mixology Whisperer' Henning Heissen, the scented cocktails at the Ritz-Carlton Berlin will be presented in a perfume bottle. The alcoholic drink will not only smell like the perfume, but its recipe will also reflect the scent in the way it tastes.

The visual menu at the bar's entrance adds a third sensory factor, giving the patrons a sense of what to expect.

4.3 Conclusions

As consumers world-wide, from many cultures, many countries, many strata of life begin to feel overwhelmed by the sheer number of marketing messages they encounter, it is important that marketers and the retail trade begin to re-think the way they engage with their customers.

Research suggests that shoppers are continually feeling a range of different emotions when they connect with brands. Feelings that may be the most effective way to make sure that customers feel positive. The strategy of sensory marketing may be one of the smartest ways for any company to trigger emotion instantly in their audience, and maintain long-term engagement. Whether because of the color of the product packaging or the signature scent in the store, the customer is more likely to remember the store, the brand, and the particular products. The trick is simply to translate the critical characteristics of the brand into compelling sensory experiences.

This chapter presents a review of an emerging application of human sensory science, namely sensory marketing, as a technique by which knowledge of sensory impressions, *i.e.*, knowledge of sensory experiences, can help sell products and services. The literature is sparse, the information hidden in the world of science articles, but is deftly applied by practitioners, and made public through stories in the popular press, rather than in the refereed scientific literature.

To put the world of sensory marketing into perspective, we may begin with the following statements.

1. A brand is one way to distinguish products from each other, while simplifying consumers' decisions in choosing an appropriate product. Brand building is not just about design but also strategy, which is even more important in the process.
2. Sensory marketing represents a way for companies to influence consumers' senses (sight, hearing, smell, touch, taste) and evoke the emotions that drive perception of the product and brand, and buying behavior.[39]
3. Consumers of today tend to take functional attributes for granted and seek an emotional aspect in the purchase process.
4. Sensory marketing creates a deeper relation between a brand and its consumers through the involvement of the human senses sight, sound, smell, taste and touch.[40]

The next logical step for the project moves to experimentation, looking at different types of behaviors, different types of shoppers, different cultures of consumers, and different natures of the *them* (who they are, what their momentary state is, such as stressed *versus* relaxed), *etc.*

References

1. E. Eriksson, A Multi-Sensory Brand-Experience: Sensorial interplay and its impact on consumers' touch behaviour (Master's thesis), Linnaeus University, 2011.
2. P. Haggard, *Nat. Rev. Neurosci.*, 2017, **18**, 196.
3. A. D. Erenkol, *J. Adm. Sci. Policy Stud.*, 2015, **3**(1), 1–26.
4. A. Krishna and N. Schwarz, *J. Consum. Psychol.*, 2014, **24**(2), 159.
5. A. Fyall, B. Garrod and Y. Wang, Experiential Marketing, in *Marketing for Tourism and Hospitality*, Routledge, 2019.
6. A. Krishna, L. Cian and N. Z. Aydınoğlu, Sensory Aspects of Package Design, *J. Retail.*, 2017, **93**, 43.
7. H. Stone, R. Bleibaum and H. Thomas, *Sensory Evaluation Practices*, Academic Press, USA, 2012.
8. D. Schmallegger and D. Carson, Blogs in Tourism: Changing Approaches to Information Exchange, *J. Vacat. Mark.*, 2008, **14**, 99–110.
9. M. Sayadi, H. Mobarakabadi and K. Hamidi, *Adv. Soc. Humanit. Manag.*, 2015, **2**(4), 100.
10. J. I. Rojas-Méndez, N. Papadopoulos and M. Alwan, *J. Prod. Brand Manage.*, 2015, **24**(1), 18.
11. E. Costell, A. Tárrega and S. Bayarri, *Chemosens. Percept.*, 2010, **3**, 42.
12. S. Bernd, *J. Mark. Manage.*, 1999, **15**(1–3), 53.
13. C. Backman, M. Crick, D. Cho-Young, M. Scharf and B. Shea, *Syst. Rev.*, 2018, **7**, 115.
14. C. R. Acevedo, V. Sérgio and L. Fairbanks, *Eur. J. Bus. Soc. Sci.*, 2017, **5**(12), 57.
15. P. Khanna and S. Mishra, *VSRD Int. J. Bus. Manage. Res.*, 2013, **3**(4), 113.
16. M. Font-i-Furnols and L. Guerrero, *Meat Sci.*, 2014, **98**(3), 361.
17. I. Berget, P. Varela and T. Næs, *Food Qual. Prefer.*, 2019, **71**, 8.
18. H. R. Moskowitz, J. H. Beckley and A. V. A. Resurreccion, *Sensory and Consumer Research in Food Product Design and Development*, John Wiley & Sons, 2007.
19. R. V. Rupini and R. Nandagopal, *J. Psychiatry*, 2015, **18**(2), 1.
20. M. Iannario, M. Manisera, D. Piccolo and P. Zuccolotto, *Adv. Data Anal. Classif.*, 2012, **6**, 303.
21. L. Y. Ljungberg and K. L. Edwards, *Mater. Des.*, 2003, **24**(7), 519.
22. P. W. Alberti, The Anatomy and Physiology of the Ear and Hearing, *Eval. Prev. Control*, 2001, 53–62.
23. H. Mcgurk and J. Macdonald, *Nature*, 1976, **264**, 746.
24. S. Gatehouse and I. Noble, *Int. J. Audiol.*, 2004, **43**(2), 85.
25. M. Engelen, *The Influence of Music on Purchasing Behavior of Consumers in a Supermarket*, (August), 2016, http://edepot.wur.nl/389468.
26. N. F. Lepora, Touch, in *Living Machines: A Handbook of Research in Biomimetic and Biohybrid Systems*, Oxford Scholarship Online, 2018.
27. V. E. Abraira and D. D. Ginty, *Neuron*, 2013, **79**(4), 618–639.
28. A. Serino and P. Haggard, *Neurosci. Biobehav. Rev.*, 2010, **34**(2), 224.

29. F. J. M. Laros and J. B. Steenkamp, *J. Bus. Res.*, 2005, **10**, 1437.
30. J. Delwiche, *Food Qual. Preference*, 2004, **15**(2), 137.
31. H. T. Lawless, *Quantitative Sensory Analysis: Psychophysics, Models and Intelligent Design*, Wiley Blackwell, 2013.
32. K. C. Hoover, *Am. J. Phys. Anthropol.*, 2010, **143**(Suppl. 51), 63.
33. R.-M. Nica, *J. Media Res. - Revista de Studii Media*, 2013, **VI3**(17), 50.
34. E. D. Adrian, *Adv. Sci.*, 1948, **4**(16), 287.
35. K. Lipartito and S. W. Mintz, *Ethnohistory*, 1998, **45**, 145.
36. S. Boesveldt and K. de Graaf, *Perception*, 2017, **46**(3–4), 307.
37. J. Kloppenburg, S. Lezberg, K. De Master, G. Stevenson and J. Hendrickson, *Hum. Organ.*, 2000, **59**(2), 177.
38. M. Macht, *Appetite*, 2008, **50**(1), 1.
39. M. Nadanyiova, J. Kliestikova and J. Kolencik, *Econ. Cult.*, 2018, **15**(1), 96.
40. T. Suhonen and J. Tengvall, Branding in the Air: A Study about the Impact of Sensory Marketing, in *Bachelor Thesis within Business Administration*, Jönköping International Business School, Jönköping University, 2009.
41. S. Harvey, *Fabrik Brands*, https://fabrikbrands.com/sensory-marketing (accessed 2020).

CHAPTER 5

Consumer Testing with Children – Challenges and Opportunities

GIANLUCA DONADINI* AND GIORGIA SPIGNO

DiSTAS – Department for Sustainable Food Process, Faculty of Agriculture, Food and Environmental Sciences, Università Cattolica del Sacro Cuore, Via Emilia Parmense 84, I-29122 Piacenza, Italy
*Email: gianluca.donadini@unicatt.it

5.1 Introduction

Globally, ~26% of the world's population is <15 years of age.[1] In Europe, which has experienced vastly decreased population growth rates, nearly one-sixth of the population is <15 years old, representing >190 million eventual consumers in that age cohort. Therefore, children are customers in one of the largest markets in many parts of the world[2] and it is crucial for the food industry to understand what drives children's food choices. Not surprisingly, children are increasingly recognized as a major market force that brings economic opportunities since they tend to influence their parents' choice decision when buying. They are key-decision makers across many product categories, having their own purchasing power, and they are future adult consumers. In addition their opinions are important elements in the product development process of foods and beverages specifically designed for this target group.[3,4] Given the size and the continuous growth of the children's food market, product development and innovation concentrate on understanding children's food attitudes, preferences and consumption. In this environment,

Consumer-based New Product Development for the Food Industry
Edited by Sebastiano Porretta, Howard Moskowitz and Attila Gere

Published by the Royal Society of Chemistry, www.rsc.org

it comes as no surprise that children are the target of intense and specialized food marketing and advertising efforts in a multitude of forms in an attempt to respond to children's demands, to build brand loyalty and to persuade them to desire a particular food product, starting when they are toddlers. For example, in light of the increasing prevalence of overweight and obesity in children, an understanding of the factors that influence eating behaviours during childhood is needed to improve the dietary patterns and health status of this age group, since children's basic biology, which favours energy dense food, does not provide an adaptive advantage in today's obesogenic environment.[5]

Consequently, children of different ages are involved in product development and healthy eating educational programs, and are exposed to media and marketing influences largely through many more diverse and integrated avenues than in the past. In particular, the direct involvement of children in sensory evaluation and consumer research has become crucial as children's sensory perceptions differ from those of adults in terms of acuity and preference,[6–8] and taste preferences are the primary predictors of children's food selection and consumption.[9,10] In this context, the methods used to explore children's behaviours in the food domain play a key role, since the direct involvement of children, and the importance of using procedures that are appropriate to the cognitive development of children, are mandatory to obtain valid and reliable results.

5.2 Main Trends in Sensory and Consumer Testing with Children

A large body of research has been conducted on sensory and consumer testing with children.[11] The main avenues of research can be identified when published papers are sorted by subject areas (Figure 5.1). Feeding behaviours, eating disorders, healthy eating and nutrition education, sensory perception and consumer science constitute the core of research with children.[12] Food safety monitoring and disease prevention, healthy eating habits, and product optimization in real eating contexts are additional areas of investigation claiming more and more attention in recent years. Trends in consumer testing with children in the last decades revealed a shift from the discovery and development of methods suitable for specific age groups of children, with regard to feeding behaviour and healthy eating habits, to the application of sensory and hedonic methods in educational interventions and in the optimization of healthier food options that are liked by children.[12]

5.3 Children as Consumers

Sensory and consumer tests for children must take into account the range of sensory and cognitive abilities of children from infancy to teen age.[13] As is known, childhood is a time when children learn many concepts and develop life-long habits. In that period, they learn about appropriate and balanced nutrition and acquire good eating habits for later years. Children grow and

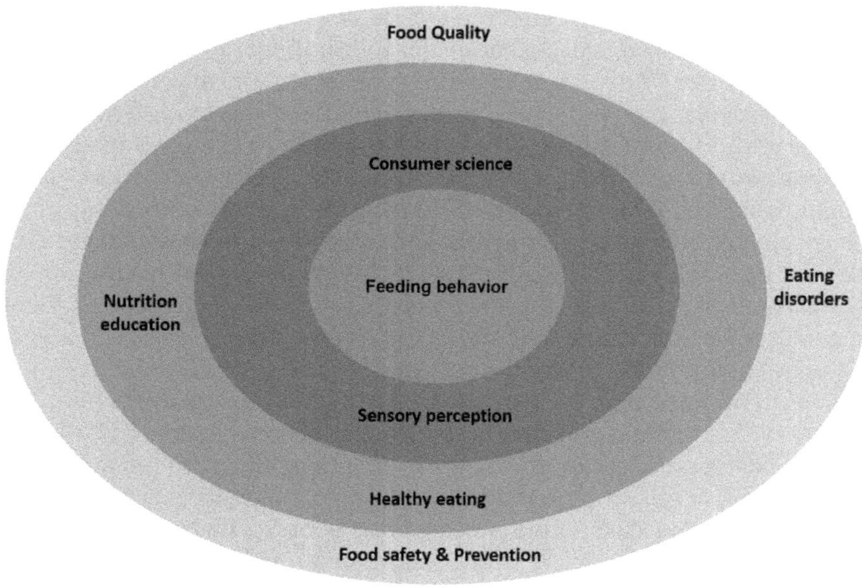

Figure 5.1 A schematic overview of the research domains of food sensory studies carried out on sensory and consumer testing with children. The closer to the centre the circle, the larger the number of papers reported by Scopus per research domain.

develop rapidly in their first 5 years of life across four main areas of development that include physical, language and communication, cognitive, and social/emotional domains. Children's development changes over time in thinking, reasoning, use of language, problem solving and disposition, learning, and their approach to interaction with their physical and social environments, which helps children think about and understand the world around them. How, when and what children learn varies largely by age since children undergo different stages to achieve complete cognitive development. According to Piaget,[14] children undergo four different stages to achieve adult thinking and knowledge. These are sensorimotor (aged 0–2 years), pre-operational (aged 2–6 years), concrete operational (aged 7–12 years) and abstract operational (≥12 years) stages, each stage encompassing numerous aspects of mental development such as reasoning, language, morals, and memory. Development involves continuous change throughout childhood, and children go through the above-mentioned stages in a sequential manner since each stage is built on the previous one and helps evolve a new level. However, the speed at which they go through them differs, as children develop at different rates in different domains rather than showing synchronous change across domains.[15,16]

Therefore, caution is needed with respect to children who are transitioning from one developmental stage to another because this age group is likely to be quite variable with respect to their cognitive abilities. That being the case, the research task and measurement technique employed in a study involving children must be age appropriate, since language, cognitive, and motor skills vary significantly by age. In the pre-operational stage, for example, children are limited in their logical thinking and are more likely to focus on a single aspect of a stimulus, whereas concrete operational children have the ability to perceive stimuli multi-dimensionally and are able to base their reaction on the simultaneous consideration of multiple aspects.[17] Very young children ~3 years of age have difficulty with concept formation (*e.g.* sweetness), classification (*e.g.* like/dislike) and ranking objects in order of magnitude.[18] They may understand the principles behind a task but their attention span may limit their ability to perform it. In addition, children have limited memory skills, which may affect their ability to remember a succession of objects or flavours presented for evaluation in a sensory test, and have trouble in ranking objects in terms of magnitude, with implications for the reliability of any scaling results from <7-year-olds.[18] Unfortunately, huge variations in skills among children of the same age are common, since each child is unique and may mature later than others do, and a child's age is far from being a perfect predictor of ability to perform specific tasks. Therefore, it is more important to consider individual skill development than chronological age when it comes to recruiting the proper sample of children to meet the test objective.

5.4 Children's Food Preferences

5.4.1 The Role of Basic Tastes and Textures

We all know that children eat the food they like and like the food that is familiar to them and it is therefore crucial for researchers to investigate patterns of food preferences among children focusing on understanding what is palatable to them.

Sensory properties influence whether a child likes a food and which food a child selects to eat, as liking a food is a matter of flavour, texture and appearance. Food preferences are the result of an interplay between genetic and environmental factors, revealing that children are suspicious and fussy about food in general and in their likes and dislikes for specific foods.[19] Genetics play a role in this regard. Some preferences are innate and moderately to highly heritable (*e.g.* preference for protein foods, vegetables and fruit). Others are learned in the course of life through eating and associating food flavours with the social context and the physiological consequences of consumption.[20–22] Fortunately, food preferences are malleable and dynamic and children are not slaves to their basic biology. Plasticity in the course of a child's life is dependent on different exposures to food so that dislikes can be reversed with simple strategies such as repeated exposure to food and modelling starting from the prenatal period. Liking of sweet taste and dislike of bitter and sour taste are inborn and universal traits of a child, these behavioural patterns in humans being consequences of evolutionary selection that favours the consumption of energy rich foods and the avoidance of bitter poisonous plants.[23]

Children have also a high liking for salty tasting foods, which, from an evolutionary point of view, would result in children preferring foods with sodium, which is needed for survival. In terms of preference, children usually prefer sweet to salty and sour foods, are neutral towards or like umami taste and do not like bitter-tasting foods.[24]

Beyond basic tastes, food texture plays a role in how foods are perceived and whether a food is liked or disliked by children. Liking of foods differing in textural properties may vary considerably among children in line with age and body development, environmental and psychological factors.[7,25,26]

The importance of texture in children's acceptance and rejection of certain food is also underlined by its association with food neophobia and pickiness that are both high during childhood.[27] Children usually like simple textures, and foods containing fewer particles are quite homogeneous and uniform and therefore can be easily manipulated in the mouth.[28,29] Segments of children with different texture preferences (hard- *versus* soft-likers) can be also identified and vary in size among children populations. Liking for different textures is culturally based[26] since children are exposed to the culinary habits of the country where they reside. In any case, individual experience with a range of foods within and outside the family is related to liking of food texture. Individual exposure to a range of textures finely-tunes culturally based hedonic responses, since children need to learn through exposure ways of managing textures in the course of their lives so as to enjoy an array of different textures, exactly as adults do.

5.4.2 The Role of Familiarity

While basic tastes such as sweetness and bitterness may be intrinsically pleasant and unpleasant, preferences for specific foods are largely learned. The diversity of world cuisines attests to the role of culture and environment in shaping what humans like to eat.[30] Exposure (once or repeated) to food plays a key role in the acquisition of food preferences, since children like familiar foods and may learn through exposure to like unfamiliar foods, which are often rejected, a culturally universal phenomenon referred to as food neophobia.[31]

5.4.3 The Role of Parents

Parents influence children's food behaviour through the products they make available as well as through the way they interact with their children.[32] In addition, parents shape children's early experiences with food and eating, providing both genes and environments for children. In fact, parents select the foods of the family diet, serve as models of eating that children learn to emulate, and use feeding practices to encourage the development of appropriate eating patterns and behaviours in children.[21] Specific behaviours or rules parents use to control what, how much, or when their child eats, being authoritative, authoritarian, indulgent or neglectful in their parenting style, have shown a stable relationship with children's eating behaviour in terms of liking, preference, choice, and intake in terms of diet quality.[33] Parents usually offer the food they like and it comes as no surprise that the mother's liking is strongly related to the child's liking. Being reluctant to serve disliked foods, parents may limit the range of flavours that children are exposed to. Therefore, parents' practices can have unintended negative effects on children's food preferences and on their self-regulation of energy intake.

5.5 How Do Children and Adults Differ in Their Preferences?

Children's sensory perceptions differ in certain respects from those of adults, and children may differ from adults in the importance they place on various product dimensions (*e.g.*, attending to appearance more than to flavour). In addition, taste preferences are not constant and vary among individuals and in each individual over time. Basic research has revealed that the chemical sensory world of children is different from that of adults, as evidenced by their heightened preferences for sweet and salty tastes.[8] For most, the preference for sweet taste remains heightened throughout childhood and does not decline to adult levels until middle to late adolescence.[23,34] Bitter perception shares several features with sweet perception. Even though people of all ages differ markedly in their perception of the same compound based on genetics, there is evidence that within the same genotype, children differ from adults in their sensitivity to some bitters that remains heightened until adolescence.[35] In addition, modulation of bitterness is based not only on bitter ligands such as sugars and salts but on the person's age, and therefore it differs in adults and children.

Notwithstanding that children prefer higher salt concentrations compared with adults, they seem to be equally sensitive to salt taste.[36] Although human new-borns are indifferent to salt taste, preference for salt emerges at ∼4–6 months of age, remains heightened throughout childhood and adolescence, and then decreases to levels resembling that of the adult during late adolescence.[37] As regards sourness, preferences are heightened during childhood and such preferences are related to children's food habits and preferences.[38]

In terms of taste perception, some studies have suggested that children have similar detection thresholds to adults, whereas other studies have suggested that children have poorer sensitivity than adults.[39] Children of different ages have good ability to identify basic tastes congruently with adults[40] but unlike adults this ability is lowered when combinations of dominant tastes occur.[41] Actually, children have reliable sensitivity in identifying basic taste in a single modality but they are poor at recognizing binary taste combinations, usually focussing on the strongest and most appealing taste of the mixture. As a result differences between children and adults seem to be greater when the recognition task increases in complexity, *e.g.*, in the case of complex stimuli such as real foods, suggesting that higher mental processing, differences in taste perception and in response strategies play a role in accounting for the age difference in children's performance.

As regards odours, children are usually less good at identifying odours than adults as they may lack odour-specific knowledge that usually accumulates over time as they grow up. Being unfamiliar to a certain extent, odours are less strongly associated with verbal descriptors, suggesting a semantic gap between children and adults that affects children's perception ability. Little is known in terms of odours' liking in children. Hoffmann *et al.*,[42] reported that children and adults may differ in preference for fruit, herbal, and spice odours, observing that children usually like sweet food odours and prefer fruit and candy odours to herbal ones.

As regards textural properties of foods, age relates to preferences, considering that oral processing varies considerably between adults and children.[26] Large differences in preferences have been observed among children of different ages, as texture perception and preference evolve over childhood in line with development of the mouth muscles, jaw, and teeth as well as innervation of taste buds. As a rule of thumb, adults prefer harder and more complex foods compared with children, although children's preferences become more adult-like with age.

Although some of the mechanisms underlying age-related changes remain unknown, functional variation in taste, smell and oral somatosensory systems may affect several sensory attributes elicited by food and play a role in acceptance of and preference for foods and flavours. The process of optimizing a product for children often results in a formulation that differs from one that is optimal for adults and from one that adults think would be optimal for children.[43] Therefore, research aimed at producing palatable food for children should directly involve children rather than extrapolating from data collected from adults. Increasing knowledge of individual variation in taste and hedonics due to both age and genetics may help food producers to achieve a more centred development of foods specifically designed for children. It is thus essential for children's products to be tested by children to obtain the appropriate direction for product development.[44]

5.6 Methods to Study Food Preferences in Children

Children can reliably perform sensory and consumer testing if age-targeted protocols are adopted and care is taken to create a friendly and inviting atmosphere with which children are comfortable. Children can participate in sensory tests concerning taste, visual appeal or texture of food, personal care products, and toys or pharmaceutical products. Children are able to perform a variety of tasks and are a valuable source of information on liking, preference, sensory perception, and emotions derived from the use of specific products. Table 5.1 provides an overview of the methods used to study food preference in children as a function of children's age.

Table 5.1 Overview of the methods most commonly used in sensory and consumer testing with children as a function of children's age.

	Newborns infants and toddlers	Pre-schoolers	Early readers	Pre-teens	Teens
Adult involvement	Primary caregiver/ trained observer/ experimenter	Adult intervention is required (familiar)	Previous and self-administered	Previous and self-administered	Not required
Evaluation techniques	Behavioural observation and diaries.	Previous plus one on one interviews Paired comparison Paired preference choice Ranking tasks (by elimination or not) Hedonic scaling (multiple point facial scale; 3-5-7-; bifurcated scales) Intensity scales Projective mapping and sorting Implicit/explicit memory	Previous plus Simple attribute rating Liking scale (pictorial or word; 9-point) Unstructured linear scale JAR scale (3-point) CATA questions Group discussion Emotion observational studies	Previous plus hedonic scales, discrimination tasks attribute scaling and ratings	All adult evaluation techniques

Logically the younger age groups are the more challenging to study in view of considerations of their language, motor skills, and social and psychological development, the appropriate methods depending on the objective of the study as well as on age (*i.e.*, cognitive and language abilities).[11]

New-borns, infants, and toddlers are not able to communicate verbally even though some verbal responses may be obtained by children approaching the second year of life. Sensory responses are usually measured individually by observing tongue movements, automic reactivity, body movements, vocalizations, facial expressions, differential ingestion, and sucking patterns that are most of the time hedonically motivated. Due to attention span limitations, the length of testing sessions and the number of products tested must be limited, and caution should be used since an observer's interpretation of children's responses might be subjective and affected by factors unrelated to the product tested.

Pre-schoolers, early readers, and pre-teens are able to perform sensory evaluation tests, such as difference or scaling tests. Paired-comparison, duo–trio and ranking tests for basic tastes, and forced-choice paired comparison are commonly used by researchers. In addition, children in this age range reliably and consistently compare and categorize foods on liking, being able to perform paired preference, preference ranking and hedonic scaling tests. Literature reports that 2–3-year-olds can reliably perform paired preference tests whereas 5-year-olds and older, who master the ability to rank things in order of magnitude,[13] can use hedonic scales of different length (3-, 5-, 7-, 9-point) that use pictures (often non-gender faces), words or a combination of picture and words depending on the child's age.[45]

When a child is illiterate, pictorial scales (Figure 5.2) are mandatory, whereas verbally anchored scales are suitable for early readers and older children, *i.e.*, ≥7 years, provided all children can read and understand the meaning of the words. These scales are similar to the traditional hedonic

Figure 5.2 An example of the non-gender facial scale used by Donadini *et al.*[59]

scales researchers use to measure the responses of adult consumers, except that they employ child-friendly anchors such as super-bad and super-good. Scale format (length, orientation, and presentation order of the labels) may bias children's use so that shorter scales with fewer choices are more likely to generate responses in the upper end of the scale range and may not be sensitive enough to represent the degree of liking by older children. A vertical scale orientation leads to positive responses whereas horizontal scales with the positive ends on the left lead to higher ratings than scales with the negative end on the left. This is because the subjects tend to choose the first alternative acceptable to them among the ordered response categories. Unlike adult consumers, children respond with the top-most category of the scale (ceiling effect), the more so the younger the children, because they lack the frame of reference that adults have developed over the years or because they are simply easier to please.

Pre-schoolers are more sensitive to demanding tasks than early readers and pre-teens. Due to relatively limited attention span, verbal, and cognitive skills, pre-schoolers can only perform simple tasks that provide quantitative data, with facial scales to measure liking, paired preference and preference with ranking techniques.

Paired-comparison and forced-choice procedures embedded or not in the context of a game are especially useful for young children because memory requirements are minimized in completing these tasks. If real foods are used as stimuli, the most reliable taste preferences are produced, but food photographs may provide a convenient alternative with adequate reliability.[46] Being able to use verbal skills, pre-schoolers may communicate their responses on one-to-one interviews with familiar experimenters. 'Ranking by elimination' and a 5-point bifurcated scale based on choice questions are suitable for determining liking among pre-schoolers.[47–49] In the former procedure, children taste a number of samples and then choose their favourite. The favourite sample is then set aside and children re-test the remaining samples and choose their favourite. This process continues until all the samples have been ranked. With the 5-point bifurcated scale children are asked if they feel the product is good or bad. Depending on the answer, they are then asked if the sample is really good (really bad) or just a little good (just a little bad). If they are not able to specify whether a food is good or bad the product is coded as neither.

Early readers develop moderate verbal skills and have increased cognitive and physical abilities compared with pre-schoolers. Facial scales, word scales and group discussions are age appropriate but attention span is still limited. Just-about-right (JAR) scale (3-point; too weak, just-about-right, too strong) may be used but only with very simple vocabulary, since scaling flavours and complex visual and textural terms may results in a high percentage of just right responses compared with older children and adults due to difficulties children have in understanding these attributes.

Pre-teens are able to complete challenging tasks and understand complex wording. Quantitative techniques such as paired comparison, paired preference choice, ranking tasks, JAR scales and hedonic scaling are suitable for this age group. One-to-one interviews are appropriate and pre-teens can participate in short interviews with unfamiliar experimenters and self-completed food questionnaires. Qualitative techniques can add in-depth information on food preferences as pre-teens provide informative descriptions about their impressions, attitudes and feeling, and use of food products. Teens are capable of completing all kinds of tests, their abilities being similar to those of adults.

As regards the use of intensity and JAR scales, literature has shown that children can generate a rich array of appropriate food descriptors and they can quite successfully scale these attributes on perceived intensity.[25,50] The evaluation of these descriptors was highly related to that of adults for appearance and texture, whereas for flavour attributes the relation was weak. In addition, children can communicate the preferred level of specific sensory attributes at different modalities, even though these sensory tasks are cognitively quite demanding, since flavour concept formation and flavour naming is a learning process that extends into adulthood. In terms of children's age, 8–12-year-olds and older children were successfully employed in descriptive analysis of foods but children as young as 4–5 years old were trained successfully to scale specific food attributes based on perceived intensity.

Qualitative methods such as focus groups and observation-based ethnography provide useful information to market researchers and product developers. These techniques help in learning opinions and needs, get closer to what children think and feel, clarify how foods are used in children's daily routine, and complement the array of information researchers acquire through quantitative investigations. Focus groups need advanced communication skills and are mostly used with pre-teens and older children, although some researchers have run focus groups with pre-schoolers. Observation-based methods, in turn, are used with children in artificial locations such as usability laboratories or in the child's own environment in the course of everyday routines. Despite the fact that conducting usability tests with children presents several challenges, there is no better way to evaluate products designed for children than when the child is involved with actual use of the product over time.

Attention has also been devoted to new approaches to existing methods that are adapted from other research areas (sorting and projective techniques, implicit memory paradigms) and applied to children [check-all-that-apply (CATA) questions].[51] Rapid methods are simple to perform and therefore promising to use in consumer studies with special populations such as children. Given the lack of scaling, CATA methods have been used with children as young as 6–9 years. These methods investigate children's perception and their drivers of liking, with the advantage that they are more flexible and less time consuming compared to traditional descriptive methods.[52–54] Under this age, vocabulary and self-expression skills undermine the validity of CATA results, which are biased by the cognitive limitations of understanding the task and do not reflect the actual perception of children.

Children from the age of 5 years can perform a structured sorting task with images and are able to complete a projective mapping task. This approach can be adapted to ask children to arrange a series of samples according to some keys properties, thus obtaining a sensory map of the attributes based on their perception.[55] Being non-verbal, projective maps are advantageous for children, especially for the youngest who may encounter difficulties in developing a sensory lexicon.[12] However, care must be taken, since with 8–12 samples simultaneously under test, fatigue and boredom may develop in children.

Wanting is an additional construct evaluated by children[56] due to its importance in food choice and consumption, and some studies have measured children's desire to eat specific foods on facial scales of different length, taking into consideration, or not, the relationship with liking. Children's age is a barrier to wanting evaluation because young children find it difficult to discriminate the concept of desire from that of liking, even though pre-schoolers as young as 4–5-years old have been employed successfully in wanting evaluation. Separation of wanting and liking tasks is suggested for young children.

Emotional response to food consumption has been evaluated by children[57] due to its importance in discriminating products with similar acceptability. Since verbal questionnaires are not suitable to use with children who have a limited vocabulary, except for pre-teens and teens, age is major barrier for children to evaluate emotions and 8–13-years of age is the cut-off in studies of this kind.[51] Emotions have been measured in ≥6-year-olds with non-verbal and non-invasive physiological measures that do not rely on the cognitive development of children. Emojis, for example, express food-elicited emotions in varied eating contexts and have the ability to describe emotional responses and discriminate between products in children.[58]

5.7 Case Studies

Two case studies are illustrated in this section to point out some specificities in conducting research with children. These studies cover a specific area of expertise in consumer science dealing with feeding behaviour and healthy eating habits of children and focus on broad educational interventions and on the optimization of healthier food options liked by children. Working with children of pre-school age is fascinating but challenging at the same time since they are developing their language skills and knowledge and have reduced ability to communicate compared with adults. Therefore, children's limited verbal skills require primarily that special consideration is given to the choice of appropriate methods of investigation. These methods often require a certain degree of adaptation to children's development. Information on the development of children's vocabulary as it pertains to the description of food attributes is crucial since it may provide an interesting perspective on similarities and differences between children and adults in the use of sensory descriptors and the importance of these attributes in the formulation of hedonic judgments.[43]

5.7.1 Case Study 1: Pre-schoolers Cheese Preference

Donadini *et al.*[59] compared six different cheeses on liking among a group of 5-year-olds ($n = 60$) with a gender split of 36 males and 24 females (Figures 5.3 and 5.4). The aim of the research was to study the preference for different cheeses among pre-schoolers and help the food service orient itself

Figure 5.3 A 5-year-old girls scoring cheese samples on liking.

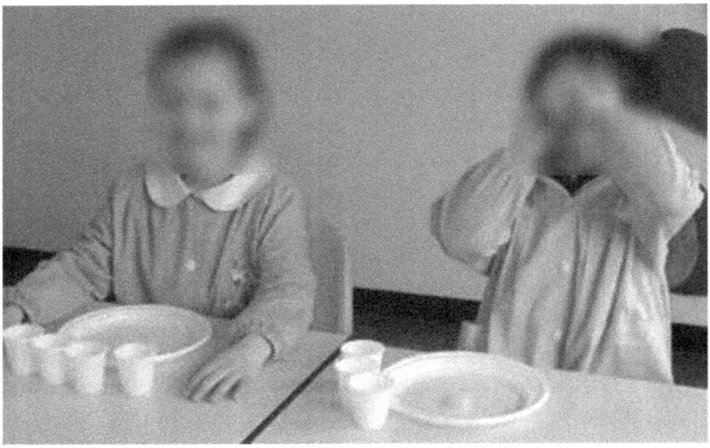

Figure 5.4 Two children participating in sensory testing in a kindergarten canteen.

in selecting highly palatable cheeses among the array of cheeses available on the market. The results showed that children discriminate the samples on liking and prefer specific cheeses to others available on the market. In addition, subgroups of children with homogeneous preference were identified and these clusters were driven by different sensory characteristics, so that nearly one child out of two focused either on flavours or texture, whereas approximately one child out of eight exclusively focused on flavours,

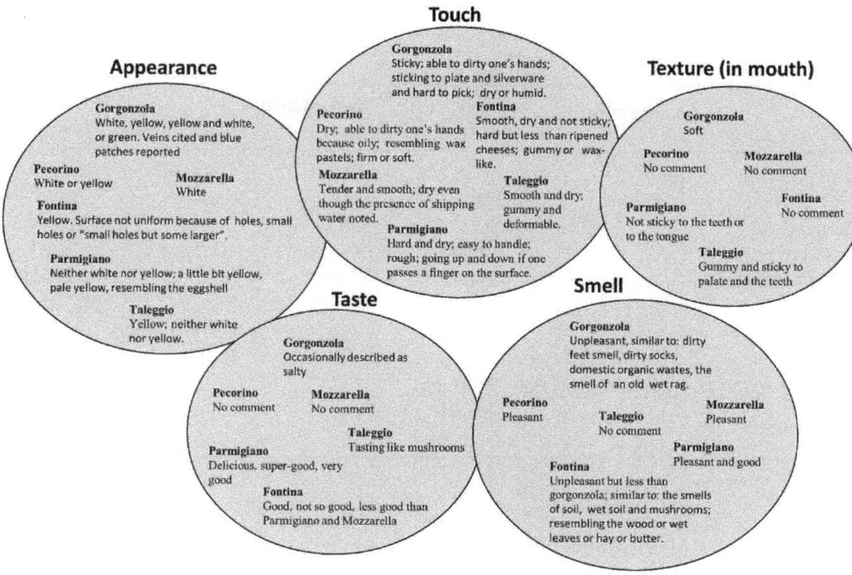

Figure 5.5 Results from focus groups with pre-schoolers on perceptions of sensory characteristics of different cheeses from Donadini *et al.*[59]

in the formulation of hedonic judgments. Besides these results, and to complement the quantitative data, two focus groups were conducted to investigate children's perceptions of the sensory characteristics of cheese. Fourteen children, equally divided between males and females, participated. The results are reported in Figure 5.5.

Children are well opinionated about the sensory characteristics of the cheeses in the test. Despite limited linguistic skills and cognitive development at 5 years of age, the children shared a large set of information with the experimenter. Interestingly, children described the cheeses along different sensory modalities ranging from visual appearance to the complex flavours. They developed a richer set of observations in two main modalities: visual appearance and touch by hands. Touch by hands in particular rarely constitutes a dimension of interest in adult consumers and attributes/descriptors pertaining to this modality are rarely used in descriptive analysis of food. The main point was that this study made it possible to sort out the sensory dimensions on which children base their focus and allowed the researcher to understand more fully the kinds of attributes that matter to children and thus identify possible avenues of product development and nutritional intervention targeting children.

5.7.2 Case Study 2: Development of Cheese Profiles

Due to limited verbal skills, attention must be paid to sensory techniques that rely largely on vocabulary generation and use. Descriptive analysis is particular challenging in this regard for children in pre-school years. In a previous study,

Table 5.2 Summary of the lexicon elicited by 5-year-olds in a previous study by Donadini *et al.* (unpublished results). The lexicon is partitioned by sensory modalities.

Visual aspect	Texture in mouth	Tastes and flavours	Smell	Touch by hand
Yellow colour	Soft	Bitter	Cheese	Soft
White colour	Hard	Salty	Fruity	Hard
Green colour	Soluble	Milk	Onion	Sticky
Brown colour	Creamy	Cheese	Garbage	Gummy
Uniform colour	Slippery	Yogurt	Stinky	Smooth
Holes	Gritty	Rotten	Hay	Gritty
	Watery	Good	Mushrooms	
Hearing	Sticky		Good	
Crackling intensity when chewing	Smooth			
	Prickly			
	Crunchy			

Donadini *et al.* (unpublished results) described the sensory profile of different cheeses by using a panel of 5-year-olds ($n = 10$) as trained judges. Results of this study will be shared and discussed in a separate communication and is beyond the scope of this chapter. However, worthy of a brief mention is the descriptive lexicon generated by children (Table 5.2) that differs from the cheese lexicon usually elicited by adults and available in the literature.

Similar to case 1, children generated a rich range of descriptors related to touch by hands that rarely constitutes a dimension of interest in adult consumers, and therefore, descriptors of this sensory modality are rarely used in descriptive analysis of food. In addition, the children noted that some cheeses crackled during mastication and that they could even hear the sensory quality of a given cheese. Therefore, unlike adults, the auditory accompaniment to eating cheese is worthy of consideration for them in the sensory evaluation of cheese. One of main points of this study is that age-appropriate descriptors may be elicited only as long as the intended consumer is employed in the lexicon-generation process. Otherwise, important sensory dimensions that matter to a specific audience may be missed. In terms of research applicability, children and adults may differ in terms of sensory perceptions and preferences and this makes working directly with children a must for successful product development, at least at some stages of product development.[60]

References

1. World Bank Group, World Population Prospects 2019, https://population.un.org/wpp/Publications/Files/WPP2019_DataBooklet.pdf.
2. R. Popper and J. J. Kroll, Conducting sensory research with children, *J. Sens. Stud.*, 2005, **20**(1), 75–87.
3. J. U. McNeal, *Kids as Customers: A Handbook of Marketing to Children*, Lexington Books, New York, 1992.

4. M. Story and S. French, Food Advertising and Marketing Directed at Children and Adolescents in the US, *Int. J. Behav. Nutr. Phys. Act.*, 2004, **1**, 3.
5. L. B. Sørensen, P. Møller, A. Flint, M. Martens and A. Raben, Effect of sensory perception of foods on appetite and food intake: a review of studies on humans, *Int. J. Obes. Relat. Metab. Disord.*, 2003, **27**(10), 1152–1166.
6. D. G. Liem, M. Mars and C. De Graaf, Sweet preferences and sugar consumption of 4- and 5-year-old children: role of parents, *Appetite*, 2004, **43**(3), 235–245.
7. L. D. Lukasewycz and J. A. Mennella, Lingual tactile acuity and food texture preferences among children and their mothers, *Food Qual. Prefer.*, 2012, **26**(1), 58–66.
8. D. G. Liem and J. A. Mennella, Heightened sour preferences during childhood, *Chem. Senses.*, 2003, **28**(2), 173–180.
9. L. L. Birch, Development of food preferences, *A. Rev. Nutr.*, 1999, **19**, 41–62.
10. A. D. McClain, C. Chappuis, S. T. Nguyen-Rodriguez, A. L. Yaroch and D. Spruijt-Metz, Psychosocial correlates of eating behavior in children and adolescents: A review, *Int. J. Behav. Nutr. Phys. Act.*, 2009, **6**, 54.
11. C. A. Forestell and J. A. Mennella, The Ontogeny of Taste Perception and Preference throughout Childhood, in *Handbook of Olfaction and Gustation*, ed. R. L. Doty, Wiley-Blackwell, 2015, 797–830.
12. M. Laureati, E. Pagliarini, T. Gallina Toschi and E. Monteleone, Research challenges and methods to study food preferences in school-aged children: A review of the last 15 years, *Food Qual. Prefer.*, 2015, **46**, 92–102.
13. J.-X. Guinard, Sensory and consumer testing with children, *Trends Food Sci. Technol.*, 2001, **11**(8), 273–283.
14. J. Piaget, Intellectual evolution from adolescence to adulthood, *Hum. Dev.*, 1972, **15**(1), 1–12.
15. K. W. Fisher and D. Bullock, Cognitive development in school-aged children: Conclusions and new directions, in *Development in Middle Childhood (70–146)*, National Academy Press, Washington D.C, 1984.
16. J. A. Cromer, A. J. Schembri, B. T. Harel and P. Maruff, The nature and rate of cognitive maturation from late childhood to adulthood, *Front. Psychol.*, 2015, **6**, 704.
17. J. H. Flavell, P. Miller and S. A. Miller, *Cognitive Development*, Prentice-Hall, Englewood Cliffs, NJ, 1993.
18. M. Gollick, Asking kids questions: Possible pitfalls. Presented at Ann. Mtg., Inst. of Food Technologists, Anaheim, Calif., June 15–19, 2002.
19. S. Anzman-Frasca, A. K. Ventura, S. Ehrenberg and K. P. Myers, Promoting healthy food preferences from the start: a narrative review of food preference learning from the prenatal period through early childhood, *Obes Rev.*, 2017, **19**(4), 576–604.

20. F. M. Breen, R. Plomin and J. Wardle, Heritability of food preferences in young children, *Physiol. Behav.*, 2006, **88**, 443–447.
21. J. F. Savage, J. O. Fisher and L. L. Birch, Parental Influence on Eating Behavior Conception to Adolescence, *J. Law Med. Ethics*, 2007, **35**(1), 22–34.
22. A. Fildes, C. H. M. van Jaarsveld, C. H. Llewellyn, A. Fisher, L. Cooke and J. Wardle, Nature and nurture in children's food preferences, *Am. J. Clin. Nutr.*, 2014, **99**, 911–917.
23. J. A. Mennella and N. K. Bobowski, The sweetness and bitterness of childhood: Insights from basic research on taste preferences, *Physiol. Behav.*, 2015, **152**(Pt B), 502–507.
24. C. Schwartz, S. Issanchou and S. Nicklaus, Developmental changes in the acceptance of the five basic tastes in the first year of life, *Br. J. Nutr.*, 2009, **102**(9), 1375–1385.
25. G. Rose, D. G. Laing, N. Oram and I. Hutchinson, Sensory profiling by children aged 6-7 and 10-11 years. 1. A descriptor approach, *Food Qual. Prefer.*, 2004, **15**(6), 585–596.
26. M. Laureati, P. Sandvik, V. L. Almli, M. Sandell, G. G. Zeinstra, L. Methven, M. Wallner, H. Jilani, B. Alfaro and C. Proserpio, Individual differences in texture preferences among European children: Development and validation of the Child Food Texture Preference Questionnaire (CFTPQ), *Food Qual. Prefer.*, 2020, **80**, 103828.
27. H. Coulthard and J. Blissett, Fruit and vegetable consumption in children and their mothers. Moderating effects of child sensory sensitivity, *Appetite*, 2009, **52**, 410–415.
28. M. Laureati, C. Cattaneo, V. Lavelli, V. Bergamaschi, P. Riso and E. Pagliarini, Application of the check-all-that-apply method (CATA) to get insights on children's drivers of liking of fiber-enriched apple purees, *J. Sens. Stud*, 2017, **32**(2), 12253.
29. G. G. Zeinstra, M. A. Koelen, F. J. Kok and C. de Graaf, The influence of preparation method on children's liking for vegetables, *Food Qual. Prefer.*, 2010, **21**(8), 906–914.
30. P. Rozin The acquisition of food habits and preferences. in *A Handbook of Health Enhancement and Disease Prevention*, ed. J. D. Matarazzo, S. M. Weiss, J. A. Herd and N. E. Miller, Wiley, New York, 1984, pp. 590–607.
31. P. Pliner, The effects of mere exposure on liking for edible substances, *Appetite*, 1982, **3**, 283–290.
32. D. Benton, Role of Parents in the Determination of the Food Preferences of Children and the Development of Obesity, *J. Obes.*, 2004, **28**(7), 858–869.
33. M. S. Faith, K. S. Scanlon, L. L. Birch, L. A. Francis and B. Sherry, Parent-Child Feeding Strategies and their Relationships to Child Eating and Weight Status, *Obes. Res.*, 2004, **12**(11), 1711–1722.
34. G. K. Beauchamp and B. J. Cowart, Development of Sweet Taste, in *Sweetness. ILSI Human Nutrition Reviews*, ed. J. Dobbing, Springer, London, 1987.

35. J. A. Mennella, M. Y. Pepino, F. F. Duke and D. L. Reed, Age modifies the genotype-phenotype relationship for the bitter receptor TAS2R38, *BMC Genet.*, 2010, **11**, 60.
36. D. G. Liem, Infants' and Children's Salt Taste Perception and Liking: A Review, *Nutrients*, 2017, **9**(9), 1011.
37. J. A. Desor, L. S. Green and O. Maller, Preference for sweet and salty in 9–15 year-old and adult humans, *Science*, 1975, **190**, 686–687.
38. D. G. Liem and J. A. Mennella, Sweet and sour preferences during childhood: role of early experiences, *Dev Psychobiol*, 2002, **41**(4), 388–395.
39. J. Blissett and A. Fogel, Intrinsic and extrinsic influences on children's acceptance of new foods, *Physiol. Behav.*, 2013, **121**, 89–95.
40. D. G. Laing, C. Segovia, T. Fark, O. N. Laing, A. L. Jinks, J. Nikolaus and T. Hummel, Tests for screening olfactory and gustatory function in school-age children, *Otolaryngol. - Head Neck Surg.*, 2008, **139**(1), 74–82.
41. E. Ervina, I. Berget, A. Nilsen and V. L. Almli, The ability of 10-11-year-old children to identify basic tastes and their liking towards unfamiliar foods, *Food Qual. Prefer.*, 2020, **83**, 103929.
42. A. C. Hoffman, R. Valdes Salgado, C. Dresler, R. Williams Faller and C. Bartlett, Flavour preferences in youth versus adults: a review, *Tob. Control*, 2016, **25**(2), ii32–ii39.
43. H. R. Moskowitz, *Food Concepts and Products: Just-in-Time Development*, Food and Nutrition Press, Trumbull, CT, 1994.
44. A. Chen, A. Resurreccion and L. Paguio, Age appropriate hedonic scales to measure food preferences of young children, *J. Sens. Stud.*, 1996, **11**(2), 141–163.
45. ASTM, *ASTM E2299 – 13: Standard Guide for Sensory Evaluation of Products by Children and Minors*, 2013, 14 pages.
46. C. A. Guthrie, L. Rapoport and J. Wardle, Young children's food preferences: a comparison of three modalities of food stimuli, *Appetite*, 2000, **35**(1), 73–77.
47. L. L. Birch, Dimensions of preschool children's food preferences, *J. Nutr. Educ.*, 1979, **11**, 91–95.
48. L. L. Birch, Effect of peer models' food choices and eating behaviors on preschoolers' food preferences, *Child Dev.*, 1980, **51**, 489–496.
49. R. Popper, M. Schraidt and J. Kroll, Testing with pre-school children: The effect of the interviewer, Presented at Ann. Mtg., Inst. of Food Technologists, Anaheim, Calif., June 15–19, 2002.
50. F. Sune, P. Lacroix and F. Huon De Kermadec, A comparison of sensory attribute use by children and experts to evaluate chocolate, *Food Qual. Prefer.*, 2002, **13**, 545–553.
51. M. Laureati and E. Pagliarini, Methods in Consumer Research, Volume 2. Alternative Approaches and Special Applications, *New Developments in Sensory and Consumer Research With Children*, Woodhead Publishing Series in Food Science, Technology and Nutrition, ch. 13, 2018, pp. 321–353.

52. H. J. Yoo, L. Machín, A. Arrua, Antunez, L. Vidal, A. Gimenez, M. R. Curutchet and G. Ares, Children and adolescents' attitudes towards sugar reduction in dairy products, *Food Res. Int.*, 2017, **94**, 108–114.
53. J. Verwaeren, X. Gellynck, S. Lagast and J. J. Schouteten, Predicting children's food choice using check-all-that-apply questions, *J. Sens. Stud.*, 2018, **34**(1), 12471.
54. M. Galler, T. Næs, V.-L. Almli and P. Varela, How children approach a CATA test influences the outcome. Insights on ticking styles from two case studies with 6–9-year old children, *Food Qual. Prefer.*, 2020, **86**, 104009.
55. P. Varela and A. Salvador, Structured sorting using pictures as a way to study nutritional and hedonic perception in children, *Food Qual. Prefer.*, 2014, **37**, 27–34.
56. D. G. Liem and L. H. Zandstra, Children's liking and wanting of snack products: Influence of shape and flavour, *Int. J. Behav. Nutr. Phys. Act.*, 2009, **6**, 38.
57. J. J. Schouteten, H. De Steur, S. Lagast, S. De Pelsmaeker and X. Gellync, Emotional and sensory profiling by children and teenagers: A case study of the check-all-that-apply method on biscuits, *J. Sens. Stud.*, 2013, **32**(1), 12249.
58. J. Sick, S. Spinelli, C. Dinella and E. Monteleone, Children's selection of emojis to express food-elicited emotions in varied eating contexts, *Food Qual. Prefer.*, 2020, **85**, 103953.
59. G. Donadini, M. D. Fumi, L. Vanoni and S. Porretta, Hedonic response to cheese in pre-schoolers, *J. Sens. Stud.*, 2012, **27**(3), 176–187.
60. S. Nicklaus, Sensory testing in new product development: working with children, in Rapid Sensory Profiling Techniques - Applications in New Product Development and Consumer Research, Woodhead Publishing Series in Food Science, *Technol. Nutr.*, 2015, 473–484.

Part Three: Concepts, Minds, Ideas and Behaviour-based Development

CHAPTER 6

Consumer-designed Features for the Labeling, Packaging, and Advertising of Insect-based Proteins. A Practical Application of the Design of Ideas

ATTILA GERE,[a] DALMA RADVANYI[b] AND HOWARD R. MOSKOWITZ*[c]

[a] Department of Postharvest and Sensory Evaluation, Sensory Laboratory, Faculty of Food Science, Szent István Uniersity, Villányi út 29-43, H-1118, Budapest, Hungary; [b] Centre for Agricultural Research, Hungarian Academy of Sciences, Plant Protection Institute, Hermann Ottó út 15, H-1022, Budapest, Hungary; [c] Mind Genomics Advisors, LLC, 11 Sherman Ave, White Plains, NY 10605, USA
*Email: mjihrm@sprynet.com

6.1 Introduction

Why eat insects?

Eating insects, or entomophagy, is an emerging field in Western countries in the past few years. Entomophagy has been part of the human diet for thousands of years.[1] Despite the frequent consumption of insects in several countries all over the world, Western consumers consider the eating of insects as a weird tradition followed in distant places. However, as the FAO report pointed out, insects may be a reasonable solution for the problems

Consumer-based New Product Development for the Food Industry
Edited by Sebastiano Porretta, Howard Moskowitz and Attila Gere

Published by the Royal Society of Chemistry, www.rsc.org

emerging from the overpopulation of the Earth.[2,3] In spite of the freshness of the topic, there are numerous publications dealing with insects as food published by food chemists, microbiologists, technologists and marketers. In the meantime, several organizations have been formed (*e.g.* North American Edible Insect Coalition, Danish Insect Network, *etc.*) and even a scientific journal, named *Journal of Insects as Food and Feed*, has been launched. The major aim of these forums is the promotion of the advantages of insects as food and feed. Several publications have identified the advantages in production (less water use, less greenhouse gas emission, less land needed compared to beef, whereas they contain more protein, vitamins and have a good fatty acid profile).[4] However, although these facts sound groundbreakingly good and one could think that they would undoubtedly convince people about the benefits of insect eating, there are some factors that inhibit the spread of entomophagy. One of the most important factors is so-called neophobia (fear of the new), more specifically, food neophobia. Several papers have identified high food neophobia scores of consumers when it comes to insect-eating.[5–7] The problem is, however, that food neophobia cannot be consciously influenced, hence it is a subconscious process and it comes to the mind so quickly that in some cases it is almost impossible to convince people even to try.[8] The subconscious nature of the problem needs to be solved by new, emerging technologies, which are able to give the answers to the question: 'How can we convince people about eating insects?' This tool is Mind Genomics, which is applied in our study to the marketing possibilities of insect-based food products.

6.1.1 Art for Art's Sake *versus* Practical Needs for Data-driven Design

At the outset, it is important for the reader to know that the author of this paper has a degree in experimental psychology, and a four-plus decades history in business as a consultant. Furthermore, it is relevant for the reader to know that the author represents a school of thought which 'looks at the world' as a set of relations between variables. With this knowledge, we can then proceed.

The notion of 'art for art's sake' comes from a school of thought in the Nineteenth Century, typified by John Ruskin. The idea is that art stands by itself, and is appreciated for itself, rather than for any utilitarian aspect that it has.

> The phrase 'art for art's sake', or l'art pour l'art, first surfaced in French literary circles in the early 19th century. In part it was a reflex of the Romantic movement's desire to detach art from the period's increasing stress on rationalism. These forces, it was believed, threatened to make art subject to demands for its utility – for usefulness of one kind or another. The phrase was taken up by writer Theophile Gautier and subsequently attracted the support of figures such as Gustave Flaubert, Stéphane Mallarmé and Charles Baudelaire. When the phrase reached Britain it became popular in the Aesthetic Movement, which encompassed painters

such as James McNeill Whistler and Lord Leighton, and writers such as Walter Pater and Oscar Wilde.

http://www.theartstory.org/definition-art-for-art.htm.

It may have been perfectly acceptable in the history of art to separate art from rationalism, and certain art from utilitarian ends. However, our increasingly technological world demands the combination of art and usefulness, in the search for better customer experience. Indeed, as people become increasingly well-off, it is the customer experience that is important, and with that change the importance of practical design, not just design for the sake of art.

Design, especially data-driven design as presented in this chapter, is the precise opposite. It is design to accomplish something.

6.1.2 Designing Foods and Other Fast-moving Consumer Goods

We don't often think of the word 'design' when we work with products, such as fast-moving consumer goods (FMCGs). We might think of design, and often do, when it comes to their packaging, but when it comes to the product itself we typically leave the specifics to professionals, so-called product developers. All too often the product developers work with marketers on the one-hand, and with so-called 'insights-professionals' on the other.

Product developers, marketers, and insights professionals come to the job of product development with different skills. Note also that they come to the job of 'product development' rather than to 'product design'. Although there is a design phase in product development, it is not design and development, but really development that is the focus in the business.

Product developers specialize in the creation of the product, whether this be the physical form, the ingredients, and so forth. Marketers specialize in the tasks needed to distribute and sell the product. Insights professionals, touching all phases of the development process, are the voice of the consumer, from identifying needs up front to understanding how the consumer reacts to the product when the prototypes are developed. It is only in packaging where the word 'design' comes in, presumably because packaging in visual, and is intimately linked with design.[9]

6.1.3 Design by Decree – The Golden Palate, the Golden Nose, and Experts 'Über Alles'

Several decades ago, during the middle of the 20th century, the notion of consumer involvement in product design was a novel idea, adopted by virtually nobody. The middle of the 20th century saw consumers as evaluators of products, but not as designers of products. The notion of 'design' was limited to the purview of experts, whether these experts were professionals like brewmasters, flavorists, perfumers, and so forth, or consumers who had undergone training to be sensory experts, or as so often happens, the spouse of those in top management. The latter, facetious as it sounds, has actually been more often than not the case in packaged goods.

The assumption during that somewhat benighted period was that consumers could not really perceive much in the way of product features. As was often the case, managers were told by their counterparts in advertising agencies that the strategy was to put out products that were reasonably acceptable, but then to create advertising that would drive the purchase and consumption of the product. The rather arrogant notion that consumers would buy most products with appropriate advertising proved true, for the first purchase, but was sadly disconfirmed when such advertising failed to generate repeat purchases because the product was simply not acceptable.

The role of the consumer in this period was relegated to evaluation, to so-called 'product tests'. The consumer might not have been thought to drive design in the up-front stages, when the decisions were being made about the 'What' of a product, but it was clear that the consumer had to approve of the final product. Thus grew the deeply held belief that the consumer could not guide, but could evaluate. An entire practice in market research grew up around that belief, the practice of 'product-testing' with consumers.[10]

6.1.4 Incorporating the Voice of the Consumer by Experimental Design

By the start of 1970, the world of product development in food and other FMCGs would change, thanks to the entrance of psychophysicists into the world of academia and business, a movement that would show the value of systematic exploration and affective design. Psychophysics is the branch of experimental psychology focusing on the relation between perception and the physical characteristics of stimuli.

Reading about psychophysics in the first part of the 20th century does not give any indication of the extraordinary importance this branch of experimental psychology would play in product design, nor could it. Psychophysics for a very long time limited itself to academic studies, first on the ability of a person to 'detect' and in turn to 'recognize' a stimulus, such as a light of a specific wavelength. This is called 'threshold' research, the study of the sensitivity of our sensory capabilities. The connection between research on sensory thresholds and the design of products is tenuous, and more truthfully, simply non-existent.

On the other hand, a new psychophysics emerged which would intimately affect the world of product design. In the mid-1950s, S.S. Stevens, a professor of psychology at Harvard, and the director of its Laboratory of Psychophysics, announced that he had discovered a profound, repeatable 'law' in psychophysics. The law was that as the intensity of a stimulus (I) increases, perceived sensory intensity (S) also increases, with the relation conforming to a power function ($S = kI^n$), and with a repeatable value of n, unique to each sensory continuum (*e.g.*, brightness, loudness, sweetness, heaviness). Each sensory continuum thus processed physical intensity in a different way, with the intensity of experience compressing the physical range (brightness, loudness), or expanding the physical range (*e.g.*, saltiness).[11]

Pure psychophysics by itself was too academic. Its application to guide product design was unclear to most scientists, perhaps because the world of the academic is to produce knowledge and further one's career in the world of learning, universities, and research laboratories. By the middle of the 1960s, two-thirds of the way into the last century, there were a fair number of psychophysicists with at least some ties to business, and an awakening by business to the possibilities of product design afforded by this new–old discipline of psychology, psychophysics.

Author H.R.M. had graduated Harvard University, a student of the aforementioned S. S. Stevens. He began work at the US Army Natick Laboratories, where the focus was a mélange of basic science and applications of psychophysics and other behavioral sciences to the needs of the US military. It was there, during the study of food preferences and menus that the application of experimental design was really born, although there were some uses of experimental design in other areas.[12]

By the time H.R.M. moved into industry, it seemed natural to apply experimental design thinking to the creation of food. The goal was to create systematically varied formulations, whose ingredients were pre-defined combinations, test these combinations among the 'target consumers', those likely to buy the product, build a mathematical model showing how the ingredients and their interactions 'drove' liking, and then maximize liking.

The notion of 'design' here is not visual design, although it could be, but rather design of components. The design is that set up to maximize the number of consumers who would be pleased with the product, and therefore buy it. Design thinking thus becomes a mathematical exercise, the dependent variable being some type of response from the person, or set of responses (pleasantness, ease of eating, appropriateness for a specific occasion). The independent variables are factors which comprise the product.

6.1.5 Moving from the Design of 'Things' to the Design of 'Ideas'

The foregoing use of experimental design dealt with physical products. At the same time, marketers were beginning to think of design to create 'ideas'. Whereas design is a positive when it comes to artistic efforts in the creation of products, when the notion of 'design of ideas' is brought up, many people who work in the creative world, the so-called 'creative class', often cringe at the thought. The idea that ideas can be designed through systematics and science can be threatening, especially to those who have the title 'creative' in the advertising and allied agencies.

Despite the fact that the 'creative class' did not embrace design of ideas, such design grew and eventually flourished. The mechanism leading to the growth was the realization by those accepting the idea that 'creation' is possibly no more than the recombination of older ideas into new blends. Once this definition of creation is accepted, then the idea of design thinking using building blocks of ideas makes a great deal of sense, and is no longer threatening.

Just as in the case of the design of products using systematically varied combinations, the design of ideas needs a mechanism by which it could work. Typically, results allay concerns, issues, and of course fears. The mechanism for design thinking of ideas was not too different from the design of products, namely a systematically varied combinations of ideas, the blends become vignettes, and are evaluated as a totality, in the same way that the blends of ingredients were evaluated as totalities.[13]

6.1.6 Enemies of Affective Design

In the world of FMCGs, the ideas behind design have taken decades to ripen and to become accepted. The delays, detours, and occasionally even full-scale blockages in the history of design come from a number of factors, including ignorance, complacency, and inertia. Each of these is, however, slowly reaching the end of its ability to block design.

1. Ignorance. The general population is aware of design for high-profile items, whether cars, electronics, home and office furnishings, and so forth. It is often design that sells. Companies in the electronics business know that they have to create new, exciting offerings, or fall behind. Witness, for example, Apple, renowned for its design. As a consequence, no one is able to ignore the principles of design, and be proud of it. It is not simply a matter of 'we've always done it this way'. Design is new, design is dynamic. Such thinking is not necessarily part of the world of FMCGs, no matter what kind of pronouncements emerge from corporations.
2. Complacency. In the world of consumer goods, success and failure occur at what often to those in electronics must seem to be a glacial pace. There is not the sense that if we mis-design the potato chip we are likely to lose out and lose our franchise, with the company going under. That unhappy state of affairs is the case with electronics. Just witness all the smartphones that are replaced each year, and the ignominious demise of the leader, Blackberry, which began the market. One doesn't see that frenetic competition in the world of foods, and certainly not in the world of snack chips. And so why invest in design? To what end? Business death, when it comes, comes slowly, gracefully, more of a silent demise in the evening than the thrashing about of one dying from a painful attack.
3. Inertia. In the world of packaged goods there is a tendency to do what has always been done before. All businesses need to know 'how they are doing', and rely upon norms. But in the world of fast moving packaged goods, the norms dictate local, specific business responses, such as increasing the advertising, or decreasing the prices in order to 'sell

more'. The goal is not to redesign, to capture the moment, to understand the underlying zeitgeist of the consumer, or if it is, then that goal is not a dominant one in the mind of the packaged-goods professional. It is incremental changes, and that phrase which is the death of innovation, 'we have to crawl before we walk, and walk before we run'. In other words, it is all about doing what has always been done, only with a slight variation.

6.1.7 Changing One's Thinking in a Connected World

Today's world is hectic, hypercompetitive, hardly stable, and all too often 'hair-raising', frightening. The old methods no longer work. The slow, majestic pace, a pace which allowed errors to be corrected readily, if not entirely prevented through careful thinking, is perceived to be 'yesterday', outmoded, and more painfully, simply irrelevant. There is no longer the dialectic between perfect and good, with the perfect taking over, becoming the enemy of the good, because time and resources allowed for it. Change is merciless.

The connectivity of people is both a blessing and a curse. It is a blessing when a group of people find a product or a service that they like. The social media thrive on positive news, positive feedback, upvotes, and the like. Recommendations are read, occasionally by those who will buy a product or a service, more often by those who have already bought the product or service. As good as positive feedback may be, negative feedback, downvotes, and cautions not to do, can be even worse.

The temptation in light of the speed and connectivity of this new century, and the reality of change as the only constant, may drive some people to avoid knowledge-building steps altogether, and just guess, just 'wing it'. Those who guess point to legends such Apple's legendary Steve Jobs, who was reported to dispense with consumer research, and simply think through the design himself. Those stories, whether true or not, give courage, foolhardy sometimes, but courage enough to guess. No one is around to record the debacles from such guessing. History is written by victors and winners, not often by losers and failures.

6.1.8 The Value of Design and the Need for Speed and Affordability to Get It Accepted

Back now to experimentation and design thinking. Design thinking is not impossible in the world of FMCGs, especially the food and beverage industry, the focus of this paper. Design thinking will not be rejected because it is new and different, at least not on the basis of that. Design thinking might be rejected on the basis of ignorance, complacency, or inertia, problems and defects inherent in the person of the business manager.

Design thinking might well be rejected on the more 'rational' basis that it is too difficult, requires too much thinking, and costs too much money, that it is appropriate, but only for those 'major projects', whose likelihood of coming about in the next year is minute, if at all. Thus, the goal is to create a system which forces design thinking as an integral part of development, a system that cannot be rejected because it has the stigma of expense, of time, and really rejected because it requires 'work'.[14]

To this end we present here the application of design thinking, using ideas, an application that can be done easily, regularly, indeed as part of the business process. The design thinking with ideas is simply the systematic mix and match of ideas according to an underlying experimental design, a set of 'recipes'. The recipes dictate what is to be combined, the features, and then how to analyze the results to understand what works, what doesn't, and what should be recombined to present a newer, better offering.

6.1.9 The Approach Illustrated by a Case History

The approach we present here is simply the systematic mixtures of ideas, of elements, of components, the evaluation of such mixtures, the identification of what 'works' in the mixtures, and the identification of different mind-sets, different criteria in the population. We do not deal with how ideas come into being, what make a good idea, and who should be doing the thinking. We are simply looking at design thinking as a strategy that can be quantified, and used easily, quickly, and inexpensively by anyone who has a smartphone and the patience to stop and think.

The approach begins with a set of questions, and ending with knowledge about what to put onto a label. The same approach can be used to design the actual parts of the label. For this paper, we focus only on the text.

Although many people believe that design is inspiration, akin to the flash of brilliance of a writer or a graphical artist, the design approach presented here is strictly systematic. The creative part occurs at the very beginning, when the raw materials are arrayed.

The process begins by identifying the topic of the project. The project must be specific, be concrete, and limited, focused, directed. By such focus, it becomes possible to shorten the process to a day or less, transforming what could be an arduous, tedious process into a dynamic, iterative chance to learn and iterate.

The topic we have chosen is the contents of a label for a new source of protein for people, insect protein. In a world with a rapidly increasing population and drain on resources, the need for sustainable protein sources has been recognized and acknowledged, sources that do not tax the planet in the way that animal proteins tax the planet.

Rationale for the project – the marching orders.

1. Help people to accept insects as food in Western countries by identifying different mind-sets.
2. Help people in need to get nutritious and quality food for an affordable price.

3. Help to protect the environment – insects have low carbon footprint.
4. Create a list of the most important product attributes affecting willingness to buy.

6.2 Materials and Methods

6.2.1 Materials

Six-by-six conjoint-type design was followed using the system IdeaMap®. The design consists of six questions (called *silos*) and each question has six answers (Table 6.1). Using the answers, so-called vignettes are created, which tell a story to the participants. A vignette comprises at most one element from each question, but several questions are absent from most of the vignettes. The question itself is never shown. Only the elements appear in the vignette.

A vignette consists of two, three or four answers to different questions, presented in different combinations to the participants. Each participant receives 48 different vignettes and rates all of them on the following two scales.

1. How likely are you to want to BUY this product if available at a good price? (1 – definitely not buy, 9 – definitely buy).
2. How do you feel when you read about this product? (1 – irritated, 2 – hopeful, 3 – curious, 4 – disgusted, 5 – happy, 6 – surprised, 7 – hungry).

After answering all the questions about the vignettes, demographic questions are presented to the participants as follows.

1. Gender
   1.1. Female
   1.2. Male
2. Place of living
   2.1. Urban
   2.2. Suburban
   2.3. Rural
3. Age
   3.1. <50 years
   3.2. >50 years
4. What level of decision-making authority do you have on purchasing food products?
   4.1. I make the final decision about the purchased food product
   4.2. I have some influence
   4.3. I have no influence
5. Marital status
   5.1. Single
   5.2. Married
   5.3. Divorced/separated

Table 6.1 The six 'questions' which constitute the silos of ideas, and the six elements given to each question.

Silo 1	**1. What should be emphasized during the marketing campaign?**
A1	Insect production has minimal carbon footprint
A2	Insect breeding creates jobs for many people all over the world
A3	Insect-based products provide quality food for people in need
A4	Insect breeding is better for the environment than livestock production
A5	Insect breeding uses far less water than livestock production
A6	Insect-based products provide high-quality nutrition (proteins, fatty acids, vitamins, *etc.*)
Silo 2	**2. What sensory expectations are present about insect-based products?**
B1	No visible evidence is insect-based
B2	Slight visible evidence is insect-based
B3	Identical taste to beef or other meats
B4	Insect ingredients limited to exotic/ethnic foods
B5	Product variety includes breads, pasta, sauce, *etc.*
B6	Only available in restaurant-prepared foods
Silo 3	**3. What nutritional characteristics would raise your interest?**
C1	Moderate consumption helps maintain healthy muscles
C2	Far less fat than meat protein
C3	Provides essential amino acids (*e.g.*, leucine)
C4	Higher quality nutrition per calorie unit
C5	Prevents disease through more unsaturated fatty acids
C6	Generally high vitamin content
Silo 4	**4. Which distribution channels should be used to reach consumers?**
D1	Available in many kinds of stores
D2	Available from local producers in small markets
D3	Available only on the internet
D4	Available in specialty foods stores
D5	Available only in the country of origin
D6	Available only in supermarkets
Silo 5	**5. What additional price would you pay for an insect-based food product?**
E1	Costs 5% more *versus* current protein
E2	Costs 10% more *versus* current protein
E3	Costs 5% less *versus* current protein
E4	Costs 10% less *versus* current protein
E5	Costs about the same as current protein
E6	Costs 15% less *versus* current protein
Silo 6	**6. What types of advertisements would raise your interest?**
F1	See TV commercial with delicious insect-based food
F2	See Facebook, Twitter, other social media ads with pictures and short comments
F3	See mobile ads with appetizing pictures and short text information
F4	Receive leaflets from supermarkets or in mail with pictures and text information
F5	Get to try insect-based products in stores
F6	Receive direct mail about insect-based products with pictures and text information

6. Income level
    - 6.1. Low
    - 6.2. Average
    - 6.3. High

7. What is your education level?
   7.1. Completed some college
   7.2. Completed some postgraduate
8. Do you follow any specific diet plans?
   8.1. No
   8.2. Yes
9. Have you heard about eating insects?
   9.1. Yes, I have heard of the eating of insects and I know what it means.
   9.2. I have heard of the eating of insects but actually don't know what it means.
   9.3. No, I have never heard of the eating of insects.

Insect eating has been associated with a high level of food and food technology neophobia. In the last two questions, food neophobia and food technology neophobia scales were presented to the participants.[15] Neophobia, the fear of anything new, can be present in several ways. In Western countries, participants often show high neophobia towards insect-eating; hence the effect of neophobia is also analyzed.

The generated study links were then sent to a random US panel of 106 respondents by an online sample provider. In order to avoid influencing of the respondents, a very brief introduction was presented, which focused more on the task they should perform and not on the topic in question. We followed the principle that all the information should come from the set of presented elements (vignettes); hence all the other influencing factors were eliminated. After reading the orientation page, participants started to answer the two questions about all 48 vignettes (Figure 6.1).

Figure 6.2 shows an example of a vignette containing three elements, each from a different silo. In other words, each expression answers one question, which together creates a 'story' or hypothetical product. No sentences are presented; Mind Genomics aims to keep the stimuli simple, only the core expressions are presented. This helps the information flow and eliminates biases coming from cumbersome composition. At the bottom of the vignette the first (rating) question is presented using radio buttons that help the participants to give the answer. Participants are encouraged not to think too much and give the response as fast as they can. We are interested in their subconscious and not conscious, over-thought answers.

As soon as the participants stated their answers, the second question appeared on the screen. Figure 6.3 shows the same vignette as Figure 6.2 but the question is about the current emotion of the participant. Again, they are encouraged to answer quickly and to not think too long. The emotion question uses seven answers while the rating question uses nine. Additionally, the feelings presented are not grouped to avoid the order effect. In some cases, participants tend to give correlated answers when the elements and/or the direction of the two scales are similar, in our case for

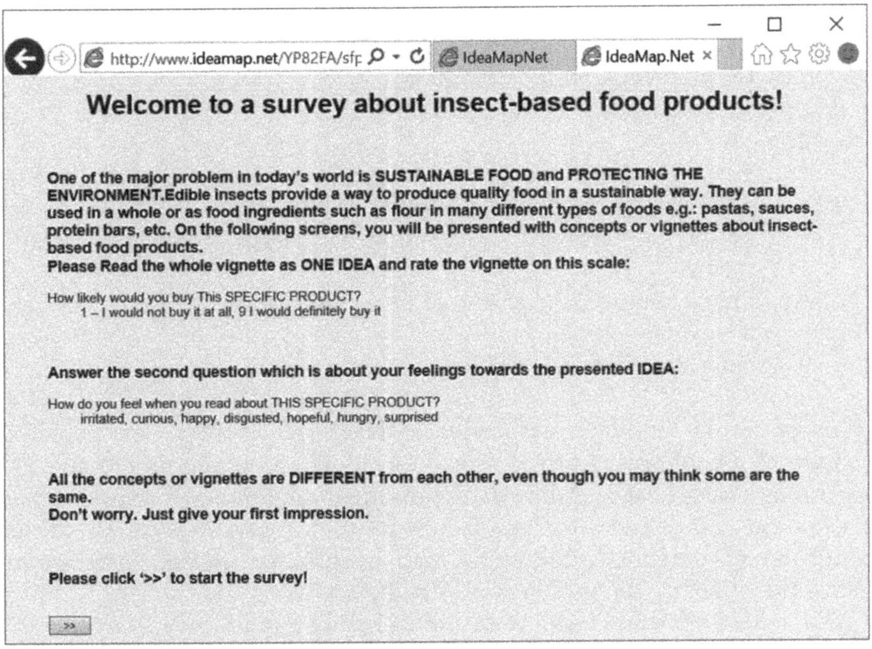

Figure 6.1 Orientation screen of the presented study. Each participant read the instructions before they started answering the questions.

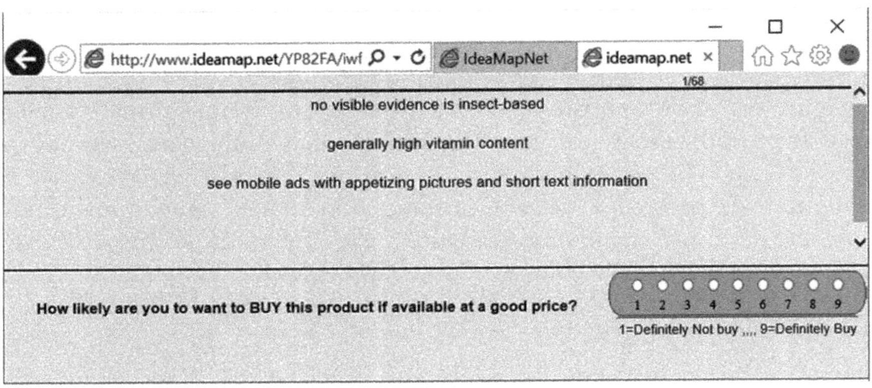

Figure 6.2 A vignette consisting of three elements, each answering one question. On the bottom of the vignette, the first rating question can be seen which focuses on willingness to buy.

example: from disgust to happy. Later, during the data analysis, the feelings will be grouped into three specific groups:

1. positive emotions: hopeful, curious, happy, hungry
2. negative emotions: irritated, surprised, disgusted

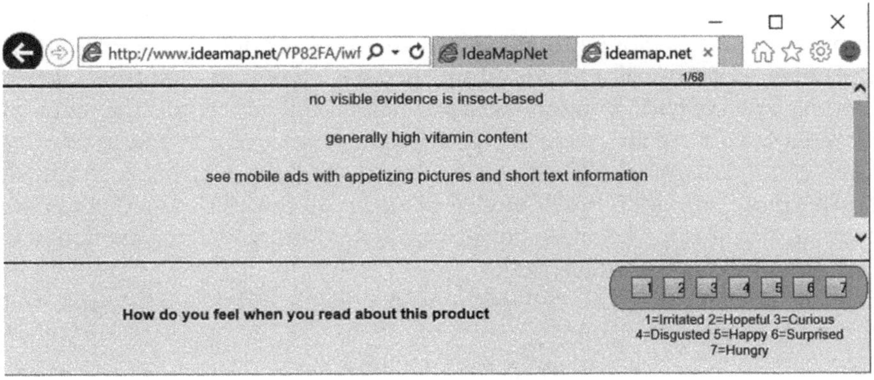

Figure 6.3 The same vignette as seen in Figure 6.2 but the question has been changed. Participants are asked to choose one feeling from the listed seven to indicate their feelings about the presented vignette.

6.2.2 Data Analysis

The final data set consisted of 36 columns indicating the elements (6×6) and two additional columns for the answers of the rating and emotion questions. Each participant has 48 rows which indicate the 48 vignettes they evaluated. The presence or absence of an element in a particular vignette (row) is indicated by 1 or 0. This means that one row of a participant contains three or four 1s because a vignette consists of three or four elements on the screen. However, each vignette has a continuous answer for the rating question (from one to nine) and a nominal answer for the emotion question. The objective of the data analysis is to identify those elements which have the highest impact on willingness to buy, or later on the chosen emotion. Several methods have been introduced to complete this regression, such as multiple linear regression analysis[16] and analysis of variance,[17] but the most widely applied method is ordinary least squares regression (OLS).[18,19] OLS regression has been shown to be superior to other commonly used partial least squares regression (PLSR).[20] Dummy variable OLS was run on the data set to assess the effects of the elements on the final score (willingness to buy) of the participants. OLS regression was run according to the following formula:

$$\hat{Y} = b_0 + b_1\mathrm{A1} + b_2\mathrm{A2} + \cdots + b_{35}\mathrm{F5} + b_{36}\mathrm{F6} + e,$$

where b_0 is the additive constant and $b_1 \ldots b_{36}$ are the coefficients of each variable in the regression model. In order to run OLS on the obtained data, transformations are needed. The so-called dummy variable approach is followed. The dummy variable approach enables us to create binary dependent variables, which means that the respondent is willing to buy the product or not. This helps the interpretation and makes it easy to compare different studies. In the first step, we transform the rating question (Question 1) to binary. This is done by assigning 0 to the values that describe willingness NOT TO BUY and 100 to those that describe willingness TO BUY.

In all cases of Question 1 (and later Question 2 also), a small random number ($<10^{-5}$) is added to all cells. A different random number is added, therefore, to each cell, ensuring that the OLS regression will 'run'. Several models are created. Out of the total panel (in which all the data are included), participants were divided by the demographic questions (*e.g.*, level of food neophobia) and a new regression was run on each subset. Furthermore, an individual model of each of the 106 participants was created. Using the coefficients of the individual models, data segmentation was done by *k*-means clustering.

In the case of emotions, two groups were created and the OLS models were run on each specific group.

(1) EMOTPOS transforms the variables to 100 if hopeful, curious, happy or hungry is selected by the participant. All the other values (emotions) are transformed to 0. EMOTPOS variable groups all the vignettes and elements that generate positive feelings in the respondents so these would be the suggested elements during a marketing campaign.
(2) EMOTNEG works similarly; in this case the values for irritated, surprised and disgusted are transformed to 100 and the others to 0. We aim to identify those elements which generate negative emotions from the respondents, hence these should be avoided.

The regression coefficients obtained are rounded in order to create more readable and understandable tables for the results. Data evaluations were done using SYSTAT version 13, from Systat Software, Inc., San Jose California USA, (www.sigmaplot.com).

6.3 Results and Discussion

6.3.1 Total Panel

In the first step, the results of all participants are evaluated and one grand OLS model is run. The coefficients of this model are presented in the third column of Table 6.2 (Total panel). As can be seen, generally low numbers are present (ranging from −3 to 3), which tell us no specific information about the panel. Higher coefficients (bold) mean that the given element raises the willingness to buy insect-based products, while lower coefficients (italic) show that participants refuse them. Columns 4–6 introduce the results of the participants divided by the question 'Have you ever heard about eating insects?', which is considered as a familiarity question. Participants had three answers to choose from based on how much they know about eating insects. The first answer 'Yes, I have heard of the eating of insects and I know what it means' lists the coefficients of those who have previous experience of entomophagy. Their coefficients are still low, close to zero, which means that their willingness would not be influenced by these elements. The second group, 'I have heard of the eating of insects but actually don't know what it means' is a promising

Table 6.2 Results are presented for all the respondents (Total panel) and divided by familiarity question and neophobia levels. Base size corresponds to the number of participants in each group, Constant shows the additive constant of the regression models, and the coefficients of each element are also shown. The table is sorted by the coefficient from question 1 (likely to buy). Strong performing positive elements (>+7) are shown in bold. Strong performing negative elements (<−7) are shown in italic.

		Total panel	Yes, I have heard of the eating of insects and I know what it means	I have heard of the eating of insects but actually don't know what it means	No, I have never heard of the eating of insects	Food technology neophobia LOW	Food technology nepophobia HIGH	Food neophobia LOW	Food neophobia HIGH
	Base size	106	63	24	19	46	60	69	37
	Constant	24	28	27	5	25	22	27	17
D1	Available in many kinds of stores	3	2	4	7	3	4	3	4
D6	Available only in supermarkets	3	1	6	5	6	0	4	1
B3	Identical taste to beef or other meats	2	3	0	1	1	2	3	0
B5	Product variety includes breads, pasta, sauce, *etc.*	2	3	−1	5	4	2	3	0
C4	Higher quality nutrition per calorie unit	2	1	0	**8**	1	3	3	0
D2	Available from local producers in small markets	2	1	2	4	3	1	1	2
A2	Insect breeding creates jobs for many people all over the world	1	2	−2	2	3	0	4	−3
A3	Insect-based products provide quality food for people in need	1	1	−5	6	2	0	3	−3

Table 6.2 (*Continued*)

	Total panel	Yes, I have heard of the eating of insects and I know what it means	I have heard of the eating of insects but actually don't know what it means	No, I have never heard of the eating of insects	Food technology neophobia LOW	Food technology nepophobia HIGH	Food neophobia LOW	Food neophobia HIGH
A6 Insect-based products provide high-quality nutrition (proteins, fatty acids, vitamins, *etc.*)	1	1	−3	**9**	2	1	3	−1
C1 Moderate consumption helps maintain healthy muscles	1	1	−1	5	2	1	3	−2
D5 Available only in the country of origin	1	1	−5	**8**	−1	3	0	2
E6 Costs 15% less *versus* current protein	1	1	−6	6	−1	2	2	−3
F1 See TV commercial with delicious insect-based food	1	1	0	2	0	1	1	−1
F2 See FacbBook, Twitter, other social media ads with pictures and short comments	1	−1	−1	**8**	2	0	3	−3
F4 Receive leaflets from supermarkets or in mail with pictures and text information	1	0	1	**4**	1	1	2	0
F5 Get to try insect-based products in stores	1	−1	−1	**8**	−1	2	1	−1

A1 Insect production has minimal carbon footprint	0	0	−5	3	0	0	1	−3
C2 Far less fat than meat protein	0	−1	0	4	1	0	1	−1
C5 Prevents disease through more unsaturated fatty acids	0	−2	0	7	0	1	2	−3
D3 Available only on the internet	0	−2	−4	7	−2	1	1	−3
D4 Available in specialty foods stores	0	0	0	−1	0	1	0	1
E5 Costs about the same as current protein	0	0	1	1	1	0	2	−3
F3 See mobile ads with appetizing pictures and short text information	0	1	−5	6	−1	1	2	−3
A4 Insect breeding is better for the environment than livestock production	−1	−1	−6	4	−3	0	−1	−2
A5 Insect breeding uses far less water than livestock production	−1	−2	−4	1	−2	0	0	−4
C6 Generally high vitamin content	−1	−3	3	3	0	−1	2	−5
E1 Costs 5% more *versus* current protein	−1	1	−9	2	0	−2	0	−4
F6 Receive direct mail about insect-based products with pictures and text information	−1	−1	−2	4	−1	0	1	−4

Table 6.2 (*Continued*)

		Total panel	Yes, I have heard of the eating of insects and I know what it means	I have heard of the eating of insects but actually don't know what it means	No, I have never heard of the eating of insects	Food technology neophobia LOW	Food technology nepophobia HIGH	Food neophobia LOW	Food neophobia HIGH
B1	No visible evidence is insect-based	−2	−1	*−5*	−2	−4	0	−1	−3
B2	Slight visible evidence is insect-based	−2	−2	*−5*	1	−4	0	−2	−2
B6	Only available in restaurant-prepared foods	−2	1	*−7*	−5	−3	−1	−2	−2
E3	Costs 5% less *versus* current protein	−2	−3	*−6*	6	−3	−1	−1	−5
E4	Costs 10% less *versus* current protein	−2	−1	*−7*	−1	−2	−2	−1	−4
B4	Insect ingredients limited to exotic/ethnic foods	−3	−2	*−7*	−3	−2	−3	−2	−4
C3	Provides essential amino acids (*e.g.*, leucine)	−3	−3	*−8*	3	−4	−2	−1	−6
E2	Costs 10% more *versus* current protein	−3	0	*−15*	3	−4	−2	−2	−4

group because their opinion is not influenced at the moment. They know the meaning of insect-eating but have no specific information about it. Their willingness is raised by the element 'available only in supermarkets' which suggests that they are open to buy these kinds of products. However, they accept only those products which have the same price as conventional meat products. They want to spend neither more nor less on insect protein. Only a low number of respondents answered that they have never heard about entomophagy but they showed positive results. The highest coefficients were obtained in this group showing that they are essentially open but need more information about insect-eating. Their willingness to buy increased when the positive nutritional effects were presented, such as 'insect-based products provide high-quality nutrition (proteins, fatty acids, vitamins, *etc.*)', or 'higher-quality nutrition per calorie unit' *etc.* They are furthermore open to advertisements through the Internet and show an attitude to try the products. These results suggest that those who do not know about insect-eating might show a positive attitude towards it.

The last four columns of Table 6.2 divide the participants by food and food technology neophobia. Participants answered the food and food technology neophobia questionnaire and their scores have been computed. Those whose final score was in the top one-third of the total score, were classified into the HIGH group, others into the LOW group. High neophobia means that the participant has negative feelings about trying new food products or products created using food technologies. Insect-eating is definitely a sensitive area for neophobic consumers, so measuring the level of neophobia helps us to understand consumers more deeply. As Table 6.2 indicates, food technology neophobia did not have a great effect on the elements: all except element 'available only in supermarkets' showed a high positive value for the LOW group. High food neophobia showed more negative coefficients: the highest value was observed for 'provides essential amino acids (*e.g.*, leucine)'. Generally speaking, evaluation of the total panel did not show specific results. In order to dig deeper into the consumers mind, segmentation was done using *k*-means clustering, in order to group people having the same way of thinking.

6.3.2 Looking for Mind-sets – Different Ways of Responding to the Same Messages (and Style Features)

Base sizes suggest that the created 4 cluster solution is well balanced, which was strengthened by cluster validity indices (data not shown). Table 6.3 presents the clusters (segments) of the data set. This grouping is different in that Table 6.2 uses *a priori* segmentation, *e.g.*, participants were grouped by a demographic variable, but Table 6.3 uses the results of Mind Genomics of each participant to create the presented segments. Each participant belongs to the segment that is the most similar to his or her way of thinking.

Segment 1: Members of the first segment have no specific opinions but they show a slight price sensitivity. Higher values were observed for 'product

Table 6.3 Detailed results of the segments created by *k*-means clustering. Italics mean strong impact.

		Segment 1	Segment 2	Segment 3	Segment 4
	Base size	31	34	15	26
	Constant	27	6	50	28
A1	Insect production has minimal carbon footprint	0	−2	−1	3
A2	Insect breeding creates jobs for many people all over the world	0	4	0	1
A3	Insect-based products provide quality food for people in need	−2	1	−3	7
A4	Insect breeding is better for the environment than livestock production	−2	−2	−1	0
A5	Insect breeding uses far less water than livestock production	−1	1	−3	−3
A6	Insect-based products provide high-quality nutrition (proteins, fatty acids, vitamins, *etc.*)	0	3	−2	2
B1	No visible evidence is insect-based	−3	−2	−3	−1
B2	Slight visible evidence is insect-based	−2	−2	−3	−2
B3	Identical taste to beef or other meats	1	4	1	−1
B4	Insect ingredients limited to exotic/ethnic foods	−1	−1	−6	−5
B5	Product variety includes breads, pasta, sauce, *etc.*	4	5	0	−3
B6	Only available in restaurant-prepared foods	−4	1	*−7*	−1
C1	Moderate consumption helps maintain healthy muscles	0	4	3	−3
C2	Far less fat than meat protein	−2	4	*−8*	2
C3	Provides essential amino acids (*e.g.*, leucine)	−2	0	*−12*	−2
C4	Higher quality nutrition per calorie unit	1	7	−1	−1
C5	Prevents disease through more unsaturated fatty acids	1	1	−2	−1
C6	Generally high vitamin content	0	3	*−7*	−4
D1	Available in many kinds of stores	3	5	1	3
D2	Available from local producers in small markets	2	4	0	−1
D3	Available only on the internet	−2	4	−2	−5
D4	Available in specialty foods stores	0	4	−5	−3
D5	Available only in the country of origin	0	3	2	−2
D6	Available only in supermarkets	2	5	−1	1
E1	Costs 5% more *versus* current protein	−3	1	3	−4
E2	Costs 10% more *versus* current protein	−6	−2	−3	1
E3	Costs 5% less *versus* current protein	−1	0	−5	−5

Table 6.3 (*Continued*)

	Segment 1	Segment 2	Segment 3	Segment 4
E4 Costs 10% less *versus* current protein	−4	2	−5	−4
E5 Costs about the same as current protein	−2	3	−1	0
E6 Costs 15% less *versus* current protein	−1	2	−4	4
F1 See TV commercial with delicious insect-based food	−1	5	−6	0
F2 See Facebook, Twitter, other social media ads with pictures and short comments	1	4	−5	0
F3 See mobile ads with appetizing pictures and short text information	−2	3	0	−1
F4 Receive leaflets from supermarkets or in mail with pictures and text information	1	2	−1	−1
F5 Get to try insect-based products in stores	−1	6	−4	−2
F6 Receive direct mail about insect-based products with pictures and text information	−1	3	*−8*	0

variety includes breads, pasta, sauce, *etc.*' and 'available in many kinds of stores', which makes them the group of Experimenters.

Segment 2: High values for product availability, nutritional benefits, and different types of advertisements were observed for the second segment, which showed the highest willingness to try insects as food products in stores. This makes them the Foodies, who like to see, buy, and taste new food products.

Segment 3: This proved to be the most negative segment, showing almost only negative coefficients, making them the group of Refusers.

Segment 4: The fourth segment showed interest in the environment and people in need, which suggests that this group would show some willingness to buy insect-based products if that helps people in need.

6.3.3 Identifying Mind-sets in the Population

Segmentation helps to define possible target groups, and if we compare the segments with the demographic data, the mind-sets can be described as presented in Table 6.4.

1. Experimenters, who accept a wide product variety available in many stores but with no visible insects, are generally females, >50 years old and completed some college.
2. Foodies have an open-minded mind-set but need to try the products first. People who care about their nutrition ready to try new foods. Females, living in suburban areas and married.

Table 6.4 Demographic patterns of the emergent segments based upon how they respond to the elements.

		Segments			
		Experimenters	Foodies	Refusers	Environmentalists
		Count	Count	Count	Count
Gender	Female	22	23	9	17
	Male	9	11	6	9
Place of living	Urban	9	7	5	9
	Suburban	11	17	7	13
	Rural	11	10	3	4
Age	<50 years	7	15	10	6
	>50 years	24	19	5	20
What level of decision-making authority do you have on purchasing food products?	I make the final decision about the purchased food product	29	25	9	20
	I have some influence	2	9	5	6
	I have no influence	0	0	1	0
Marital status	Single	8	11	3	11
	Married	13	16	5	12
	Divorced/separated	10	7	7	3
Income level	Low	20	23	9	16
	Average	4	6	1	9
	High	7	5	5	1
What is your education level?	Completed some college	23	24	7	8
	Completed some postgraduate	8	10	8	18
Do you follow any specific diet plans?	No	22	23	9	14
	Yes	9	11	6	12
Have you heard about eating insects?	Yes, I have heard of the eating of insects and I know what it means.	15	22	8	18
	I have heard of the eating of insects but actually don't know what it means.	8	10	3	3
	No, I have never heard of the eating of insects.	8	2	4	5

3. Refusers are the most negative group, they are ready to pay 5% more if it helps maintain healthy muscles. They live in urban/suburban areas, divorced or separated.
4. Environmentalists appreciate the positive effects of insect-based products on the environment and on people in need but want to pay less for the products. The majority of the group is female, living in suburban areas and single or married.

6.3.4 Emotions about Eating Insects

Table 6.5 lists the coefficients obtained after grouping the results by emotions. Respondents felt irritated when reading elements about higher prices or availability on the internet. Surprising elements were those about fat

Table 6.5 Results of the second question about emotions.

How do you feel when you read these vignettes?	Irritated	Curious	Disgusted	Surprised
Sorted by Irritated				
Available only on the internet	**5**	8	7	1
Costs 10% more *versus* current protein	**5**	1	11	3
Insect-based products provide high-quality nutrition (proteins, fatty acids, vitamins, *etc.*)	**4**	5	9	4
Costs 5% more *versus* current protein	**4**	4	7	4
Receive direct mail about insect-based products with pictures and text information	**4**	10	7	3
Sorted by Curious				
See mobile ads with appetizing pictures and short text information	2	**11**	6	2
Only available in restaurant-prepared foods	1	**10**	9	3
Receive direct mail about insect-based products with pictures and text information	4	**10**	7	3
Available in specialty foods stores	3	**9**	6	2
No visible evidence is insect-based	2	**9**	7	4
Sorted by Disgusted				
Slight visible evidence is insect-based	2	4	**14**	3
Costs 10% more *versus* current protein	5	1	**11**	3
Insect-based products provide quality food for people in need	4	4	**9**	2
Costs 15% less *versus* current protein	2	5	**9**	3
Only available in restaurant prepared foods	1	10	**9**	3
Sorted by Surprised				
Far less fat than meat protein	2	7	6	**6**
Identical taste to beef or other meats	2	6	5	**6**
Insect breeding creates jobs for many people all over the world	3	5	8	**5**
See Facebook, Twitter, other social media ads with pictures and short comments	3	9	6	**5**
See TV commercial with delicious insect-based food	3	8	9	**5**

content and that insect protein has a taste identical to beef, which suggests that these elements gave new information to the respondents. Putting more emphasis on this information may result in higher willingness to buy insect-based products. The most intense emotions were curious and disgusted. Lack of visibility enhanced curiosity as well as availability in restaurants as prepared food. High curiosity values for these elements show that presenting food in restaurants in a way that the insect is not seen, could help raise the willingness to try insect-based food products. An intense 'yuck' factor is present when participants rate a vignette as disgusting when elements about higher price and visible insect are present.

6.4 Discussion – Data, the 'Facts of the Experiment'

A novel approach in entomophagy was introduced in the present paper, focusing on the marketing possibilities of insect-based food products. Mind Genomics addresses the research question from an experimental design point of view, it creates an incomplete design of 6 factors (questions or silos) and their 6 levels (answers or elements).

The results obtained showed that four distinct segments (mind-sets) can be differentiated. Experimenters, Foodies, Refusers and Environmentalists should be addressed in different ways; however, their socio-demographic background is not significantly different. Our results strengthened the idea that the major barriers of entomophagy are disgust and neophobia, but there are some promising points also.

The high coefficients of surprise show that some elements are new and interesting to people. These elements were about the nutritional and environmental benefits of insect eating, and the meat-like taste. Results suggest that there is a significant group in the sample who are open to insect eating but would need further information to make their final decision about the topic.

The study presented draws attention to the fact that conventional methods may not be able to uncover hidden preferences, *i.e.*, segments, present in the big data. Measuring socio-demographic factors is not enough to explain the behavior of consumers. The results presented showed that people having similar socio-demographic background (*e.g.*, age, gender, income level, *etc.*) may have significantly different mind-sets.

Methods that are able to measure subconscious behavior or emotions are becoming more and more important in order to uncover the true mind of consumers. Mind Genomics proved to be an excellent tool to measure the factors influencing the willingness to buy insect-based protein. The four mind-sets uncovered show us that different people should be addressed with different messages. The demographic background of the mind-sets is not significantly different, the way their members think about insect-eating differs significantly between segments.

However, identifying the mind-set of a group of participants on a given topic (*e.g.*, insect eating) does not mean that the same group exists in other topics (*e.g.*, novel food technologies). In order to get a clear picture about the whole mind of the participants, several smaller studies are needed which will result a series of mind-set memberships, describing all the participants.

As has been shown, this information can be easily extended to emotions. Certain elements, or combination of elements, elicit given emotions, such as visible insects raise disgust. Having the mind genomic of the participants enables the researchers to focus on those things that are important for the people, even if they are not aware of it.

6.5 Discussion and Conclusion – Solving the Problem of Affective Design by Experiment and the Creation of Structured, Small Databases

Creating a new product by finding out what people feel to be interesting in the product has typically been done by consumer researchers on the one hand, and product developers on the other. The former uses questionnaires, and observes what happens in the market. The latter uses intuition, feeling, experience, a sense of what people 'want'.

We propose that for the development of products based on what people like (rather than need), that there be introduced into the development process a step called *affective experimentation*, as we demonstrate here. The notion is that through small experiments that either describe the product or show the product, when the test stimulus varies systematically, one can infer causation, *i.e.*, one can identify the driving features of the product, or in this case the message about the product.

We don't hear very much about experimentation and affect. There is always the interplay between the artist and the researcher. Experiments are left to the more objective aspects of a product, the physical performance, the nutrition, and so forth. Design is typically left to the creative, with the consumer providing either very early input in focus groups, or simply a go/no-go vote at the end of the development process. We propose a whole new stage, somewhere in between, to test the affective value of the parts of the offering, doing so by small, simple to run, simple to analyze experiments.

The notion of the small experiment to understand causation of likes/dislikes provides an entirely new dimension for affective design. The experiments weave a potentially productive, time-shortening path encompassing the artist/designer 'who knows', the scientist who 'creates', the researcher who 'measures and hypothesizes why', and the consumer who is the ultimate judge of that which is created.

The future of small, just-right 'experiments' in a world of big art, big science, big data will become brighter as the world of data gets larger, the world of competition becomes increasingly fierce, and the ability to do small, iterative experiments emerges with technology, technology which creates the stimuli, and technology which involves the consumer. We don't know when those futures and worlds will converge, but it is becoming increasingly clear from the news that the fortunes of companies are always at stake, and are no longer permanently assured by their successes in the past.

Acknowledgements

A.G. is grateful for the support of the Premium Postdoctoral Researcher Program of Hungarian Academy of Sciences and the National Research, Development and Innovation Office of Hungary (OTKA, contract No K119269). The project was supported by the European Union and co-financed by the European Social Fund (grant agreement no. EFOP-3.6.3-VEKOP-16-2017-00005).

References

1. A. van Huis, *J. Insects as Food Feed.*, 2017, **3**(3), 161.
2. A. van Huis, J. Van Itterbeeck, H. Klunder, E. Mertens, A. Halloran, G. Muir and P. Vantomme, *Edible Insects. Future Prospects for Food and Feed Security*, Rome. https://doi.org/978-92-5-107595-1, 2013.
3. A. van Huis, *Annu. Rev. Entomol.*, 2013, **58**, 563.
4. R. T. Gahukar, Chapter 4 – *Edible Insects Farming: Efficiency and Impact on Family Livelihood, Food Security, and Environment Compared with Livestock and Crops*, in *Insects as Sustainable Food Ingredients*, ed. A. T. Dossey, J. A. Morales-Ramos and M. Guadalupe Rojas, Academic Press, San Diego. https://doi.org/10.1016/B978-0-12-802856-8.00004-1, 2016.
5. R. Caparros Megido, C. Gierts, C. Blecker, Y. Brostaux, E. Haubruge, T. Alabi and F. Francis, *Appetite*, 2015, **52**, 237.
6. C. Hartmann, J. Shi, A. Giusto and M. Siegrist, *Food Qual Prefer.*, 2015, **44**, 148.
7. S. Piha, T. Pohjanheimo, A. Lähteenmäki-Uutela, Z. Křečková and T. Otterbring, *Food Qual Prefer.*, 2018, **70**, 1.
8. A. Gere, G. Székely, S. Kovács, Z. Kóka and L. Sipos, *Food Qual Prefer.*, 2017, **59**, 81.
9. H. Traitler, Innovation Partnerships as a Vehicle toward Open Innovation and Open Business, in *An Integrated Approach to New Food Product Development*, ed. H. R. Moskowitz, I. Sam Saguy and T. Straus, CRC Press LLC, Boca Raton, vol. 77, 2009.
10. H. R. Moskowitz, J. H. Beckley and A. V. A. Resurreccion, Innovation's Friend: Integrated Market and Sensory Input for Food Product Design and Development, in *Sensory and Consumer Research in Food Product Design and Development*, Wiley-Blackwell, 2012, pp. 83–114. https://doi.org/10.1002/9781119945970.ch4.
11. P. Killeen, *J. Exp. Anal. Behav.*, 1976, **25**(1), 123.
12. R. M. May, *Nature*, 1976, **261**, 459.
13. V. Milutinovic and J. Salo, *Mind Genomics – A Guide to Data-driven Marketing Strategy*. Springer International Publishing. https://doi.org/10.1007/978-3-319-39733-7, 2016.
14. D. C. Chou, *Comput. Stand. Inter.*, 2018, **55**, 73.

15. W. Verbeke, *Food Qual. Prefer.*, 2015, **39**, 147.
16. T. Lima Filho, S. M. Della Lucia, R. M. Lima and V. P. R. Minim, *Food Res. Int.*, 2015, **72**, 126.
17. J. C. de Andrade, É. S. Nalério, C. Giongo, M. D. de Barcellos, G. Ares and R. Deliza, *Food Qual. Prefer.*, 2016, **53**, 168.
18. L. C. Shan, A. De Brún, M. Henchion, C. Li, C. Murrin, P. G. Wall and F. J. Monahan, *Meat Sci.*, 2017, **131**, 82.
19. A. Annunziata and R. Vecchio, *Food Qual. Prefer.*, 2013, **28**(1), 348.
20. S. R. Jaeger, L. H. Mielby, H. Heymann, Y. Jia and M. B. Frøst, *J. Sci. Food Agric.*, 2013, **93**(15), 3682.

CHAPTER 7

Artificial Intelligence to Identify Ideas

SOMSUBHRA GANCHOUDHURI,[a]
SOMRITA GANCHOUDHURI[b] AND HIMANSHU UPRETI*[c]

[a] Cofounder & CEO, Ai Palette, 75 Ayer Rajah Crescent, Singapore; [b] National University of Singapore, Singapore; [c] Cofounder & CTO, Ai Palette, 75 Ayer Rajah Crescent, Singapore
*Email: himanshu@aipalette.com

7.1 Introduction

Increasingly, Artificial Intelligence more commonly referred to as AI has become the buzzword in industry. With the emergence of the Internet and realization by scientists in the 1940s of the possibility of using digital computers to solve complex problems,[1] there has been tremendous work in this field and today we use AI algorithms to predict problems and provide solutions. AI is now portrayed as a solution provider. Once considered to be an esoteric topic for being too math-heavy, AI has now become a household term with its presence in advertisements for commonly used items such as smartphones.[2]

7.2 About AI

Several scholars highlight the various stops and starts of AI''s journey over the years.[3,4] They cite that the primary reason for the negative publicity of AI overtime was a lack of understanding of AI, which led to the formulation of false hopes about its the accomplishments. AI in simple words refers to a set of algorithms that analyses information, mainly data, for decision-making and the performing of tasks.[2] AI is concerned with intelligent behavior in

Consumer-based New Product Development for the Food Industry
Edited by Sebastiano Porretta, Howard Moskowitz and Attila Gere

Published by the Royal Society of Chemistry, www.rsc.org

artifacts. Intelligent behavior, in turn, involves perception, reasoning, learning, communication, and acting in complex environments. AI has as one of its long-term goals the development of machines that can do these things as well as humans can, or possibly even better. Another goal of AI is to understand this kind of behavior, whether it occurs in machines or in humans or other animals. Thus, AI has both engineering and scientific goals.[1] However, research stresses that the algorithms designed to achieve the desired goals do not necessarily need to be human goals or methods of achieving those goals. There are four ways to define AI: (i) acting humanly, (ii) thinking humanly, (iii) thinking rationally, and (iv) acting rationally.[4]

AI devices fall into two fundamental categories – applied and general. Applied AI which is common in nature refers to systems designed to intelligently trade stocks and shares; whereas generalized AI, less common in nature, refers to systems that are capable of handling any task.[5] Generalized AI, an emerging field, aims at functioning as a human with an ability to deal with complex situations with analytical thinking and depth of knowledge so that it can deal with diverse topics. However, many experts claim this to be elusive, stating that it hasn't yet reached the intelligence of a 6-year-old. Applied AI is of great significance today through enabling a high-functioning system of replication and providing valuable information and services to users in real-time.

The ways to classify AI are often arbitrary. AI is often classified as strong AI – having a broad design, and weak AI – having a narrow design. The problem with the broad design is that it tends to become less accurate, whereas AI with a narrow design tends to be too restrictive.[4,6]

According to Hintze,[7] AI can be classified into four different types that highlight the barriers that separate AI from human beings.

Type I AI. Reactive machines: this type of AI is reactive in nature and is incapable of storing any memory or using past experiences to make informed decisions, *e.g.*, Deep Blue, IBM's chess-playing supercomputer that acts on the current moves of its opponents and its predictions of their next moves. These reactive machines are the simplest forms of AI and the most dependable, since they are programmed to act in a similar way every time they face a similar situation. They interact with the world the way they sees it without having any internal concept of the world.

Type II AI. Limited memory: this type II machine contains memories that are transient in nature. For example, self-driving cars have preprogrammed memories that help them to navigate when there is a red light or when there is a need to change lanes. However, it is important to note that such information is not stored in the car's library of experience so that it can learn from it over the years, in the way that humans learn from their driving experience.

Type III AI. Theory of mind: in psychology this refers to the presence of individual thoughts and emotions that impact one's behavior.

The addition of this vital component in AI machines is considered to be important if AI machines are to be on a par with humans.

Type IV AI. Self-awareness: this is considered as the last step of development of AI where machines will have a consciousness of their own. This is an extension of the theory of mind.

7.2.1 History of AI

It was in 1956 when John McCarthy first coined the term 'Artificial Intelligence' during the first academic conference on AI at Dartmouth College, Hanover, New Hampshire. Although this term was officially coined during this time, long before that classical philosophers conceptualised human thinking as a symbolic system. Historically, there exists discourse on the possibility of inanimate objects coming to life as intelligent beings. For example, ancient Greek mythology had discussions on the possibility of the existence of robots, and Chinese and Egyptian engineers constructed automatons.[8] Since it was officially founded, AI has gone through numerous waves of ups and downs. There have been periods referred to as AI winter that witnessed reduced government funding along with an overall drop in interest in the field. AI winters occurred when AI failed to manage the expectations highlighted by proponents. AI's journey is also often referred to as a roller coaster ride due to its various successes and setbacks. Below are some of the highlights of the transformation of this field since 1952.[3]

- 1954–1974: this period witnessed an increase in the algorithm toolkit along with the flow of funds.
- 1982–1990: the Japanese government invested $400 million on the Fifth Generation Computer Project (FGCP), however, this produced unsatisfactory results.
- 1990–2000: IBM's Deep Blue defeated the grandmaster of chess, Gary Kasparov.

At present, there is AI hype all around and machine learning is viewed as the reason for this. Machine learning is a technology that helps computers to learn from data.[4]

7.3 AI for Identifying Ideas

The human brain identifies ideas based on the information or data point that it has in front of it, a process that it has learned over the years. Similarly AI algorithms can also be trained to identify ideas based on the same principles as the human brain would do. Here are the key steps to having an AI model that can identify ideas.

7.3.1 Data Sources

Data is very important for training any AI model, the bigger the data set the better AI models works. Today we have a huge amount of data around us. A large of amount of this data is UGC (user generated content) across online channels such as social media, retail, blogs and forums, which provide a rich dataset with millions of data points that can be leveraged for training AI algorithms to provide predictive insights. However, the challenge here is to identify which data are important and filter out the irrelevant data.

7.3.2 Identifying Dominant, Emerging and Growing Ideas

Once you have data over a substantial period of time, you can then identify patterns emerging; and on that basis you can identify which of these ideas are dominant and have existed for a longer period of time. Growing ideas are the ones that have seen sustained positive growth in the recent past. Emerging ideas are the new ideas that have recently come up and could become a growing or dominant trend tomorrow, *e.g.*, gut health in food.

The fading ideas are the ones that have started to lose interest and are on their downward negative growth; pursuing these ideas should be avoided.

Similarly new emerging ideas and growing ideas can be identified.

The historical patterns can then be used for predicting future trajectories of these ideas.

7.3.3 Identifying the Context of the Ideas

The other important aspect wherein AI algorithms play a big role is identification of ideas in a particular context. For example, a food-trained AI algorithm can predict the next big dish or flavor idea. Similarly, an AI algorithm trained on finance data can predict which stock will go up and down in the market.

The context brings in a degree of efficiency when it comes to the output of predictive AI algorithms. The context can further be classified to understand better why a particular idea is gaining popularity; this insight also helps in creating a new product/service based on that idea.

7.3.4 Predicting Future Growth

Predictive AI algorithms can potentially identify emerging and white space opportunities by going through a large dataset and then predicting the ideas that will achieve future growth.

By training the model on big data sets the algorithms are able to predict, with a reasonable degree of accuracy, what the future growth of the ideas is going to be.

7.3.5 Prioritising Ideas

AI can be used for funneling down the top ideas from a huge set of ideas. The algorithm trained on previous data can identify the patterns that cause an idea to be successful and then potentially apply those principles to a 'new' idea and predict its future degree of success.

This decision is made based on multiple factors, the self-learning algorithm over a period adjusts the weightage given to all these factors to make the predictions better over time.

7.4 Future

We should expect AI and machine learning to become part of companies' strategy, and they will expect it to create competitive advantage for them. Future decisions will become heavily dependent on data analytics. Companies who are able to do it faster will be leading the pack whereas the others will find themselves on the sidelines.

The corporate customer departments will undergo a transformation. We will see data scientists working closely with other regular functions; however it is unrealistic to imagine that occupations will become completely automated. AI will only augment our intelligence to identify the best ideas with an accuracy never seen before.

References

1. B. J. Copeland, Retrieved from https://www.britannica.com/technology/artificial-intelligence, 2020.
2. Z. Voulgaris and Y. E. Bulut, *AI for Data Science: Artificial Intelligence Frameworks and Functionality for Deep Learning, Optimization, and Beyond*, Technics Publications, 2018.
3. R. Anyoha, Retrieved from http://sitn.hms.harvard.edu/flash/2017/history-artificial-intelligence/, 2017.
4. J. P. Mueller and L. Massaron, *Artificial Intelligence for Dummies*, ISBN: 978-1-119-46765-6, 2018.
5. B. Marr, Retrieved from https://www.forbes.com/sites/bernardmarr/2016/12/06/what-is-the-difference-between-artificial-intelligence-and-machine-learning/#5e2dd5112742, 2016.
6. R. Young, Retrieved from https://itchronicles.com/technology/applied-generalized-artificial-intelligence/.
7. A. Hintze, Retrieved from https://theconversation.com/understanding-the-four-types-of-ai-from-reactive-robots-to-self-aware-beings-67616.
8. T. Lewis, Retrieved from https://www.livescience.com/49007-history-of-artificial-intelligence.html, 2014.

CHAPTER 8

Mind Genomics (BimiLeap) to Create New Ideas

NAMRATA MEHTA-SHAH,[a] SHRADHA MEHTA[a] AND RYAN ZEMEL[b]

[a] Coup D'ePop, USA; [b] LimbicReviews, USA

The year is 2020. TikTok and Zoom calls are 'all the rage', the fate of our world relies on social distancing to #flattentheCo-vidcurve and veganism is at its all-time high after Joaquin Phoenix's Oscar's acceptance speech. It is no surprise that wholesome and natural food is the hottest commodity right now, whether it is the newest alternative milk on the market or the unexpected better-for-you snacking option like seaweed chips or cauliflower pretzels. Oh, and roasted water lily seeds, of course.

When the universe revolves around food consumption and every corner we turn is driven by what we eat, new-to-market foods are extremely popular and are driving people to innovate within the food industry. This holds particularly true for better-for-you food trends. Think, health-driven; think, the next superfood. Often we rely on Whole Foods and Trader Joe's to introduce us to the newest fad in food, but sometimes the greatest ideas are right in front of your eyes. All it takes is a eureka moment to realize the potential it can have if put forth into the universe. The key example of this chapter is: roasted water lily seeds, culturally known as 'Makhana' in Southeast Asia. These seeds have been the Eastern world's best kept secret until now (Figure 8.1).

Originating in India, water lily seeds have been a go-to ingredient in Indian cooking and more popularly eaten as a roasted snack flavored with Indian spices and herbs. Available in food stalls, groceries, theaters and

Consumer-based New Product Development for the Food Industry
Edited by Sebastiano Porretta, Howard Moskowitz and Attila Gere

Published by the Royal Society of Chemistry, www.rsc.org

Figure 8.1 Water Lily exhibit, New York Botanical Gardens, New York, NY August 2019.

more – its reach spans into the masses and it is liked by all: young and old, rural and urban, those homebound and working professionals. Water lily seeds, widely known as 'Makhana' or 'Phool Makhana' in South East Asia, derive from water lily plants grown across water ecosystems in Bihar, India.

Nutritious shells containing water lily seeds are sustainably harvested from water lily plants by local farm collectives. Seeds are then hammered out of the shells and still, to this day, hand-popped over high heat. Through a roasting and flavoring process, water lily seeds are ready for consumption. Packed with protein, calcium, potassium, magnesium, and iron, its properties aid in anti-inflammation, anti-aging, maintenance of a low glycemic index, while also qualifying as an 8 out of 8 allergen-free seed. Its health benefits have been boasted about for centuries and passed down through generations by a population of 1.1 billion people. Seeing the potential for it to explode into the global healthy snacking market was a no brainer.

Let's back up and talk healthy snacking for a minute, and how it has exploded into mainstream consumption as a lifestyle staple. 'Vegan', 'non-GMO', 'gluten free', 'pure ingredients' are just some buzz words that define how people consume food nowadays. Consciousness and awareness drive lifestyle changes when it comes to health and self-care. More and more people are investing time and effort to make sure their diets are nutritionally balanced with clean wholesome eating. Rather than go for the fully processed junk, people opt for snacks with ingredients that they can pronounce with an ingredients list that is less than a novel. No wonder the healthy snacking industry is set to grow three times faster than all other snacking markets. With a projected growth of 47% in the US market from 2017 to 2025, with seeds and nuts being the biggest contributor, new-to-market projects will race to make it onto prime shelf space. Digesting this information and further analyzing our research, all whilst munching away on roasted water lily seeds, it was only a matter of time before we recognized an opportunity and became excited to delve further into the ideation process.

Enter, Coup D'epop: a revolutionary healthy snack of roasted water lily seeds founded by two avid foodies hoping to cause uproar in the healthy snacking space with the goal to #strivetoeatbetter.

Very early on, we understood the struggles of introducing a product of this kind into a market that was oblivious to its existence. The first hurdle would be education. How were we going to spread awareness of a product that people had no prior knowledge of, let alone tasted? How were we going to get the consumer to try it? And once tried, would they accept it and consume it regularly? This was our ultimate goal, to be so revolutionary that it rooted itself into everyone's daily lifestyle, be that at school, on the move, at work, after the gym, or even just at home. Trust me, we get it. Who wants to put something random they have never seen or heard of before in their mouth without knowing what it really is. That is why it was imperative to find a familiarizing point of sale that consumers could relate this product to.

In seeing how roasted water lily seeds were being eaten and paying close attention to aesthetics, texture, crunch profile, as some factors, there was a direct correlation between our seeds and popcorn. Understanding popcorn's relatability to roasted water lily seeds fueled our excitement. In seeing the opportunity that lay in front of us to take over popcorn's existing market share with our relevant product, our project, which started as an idea, was slowly taking form as a business plan with a vision and a purpose.

Our journey began with creating our product ideation, or at least the first wave of our product ideation. This was prior to our popcorn comparison realization. The thing is, brand identity and an original narrative rely deeply on execution. As founders, we knew what we wanted the brand to represent and the emotions we wanted conveyed. We wanted the brand to be quirky, fun, bold and approachable. We wanted to be the hot new kid on the (*whole and natural eating*) block. We wanted our name to be a strong one, bonus points for something 'punny', and we wanted the consumer to get it – get all of it at the first introduction.

Naturally, as all well-educated startup mavens would do – we turned to Google and designer forums and found a couple of designers who freelanced on product packaging. We created a brief with words and colors and industry statistics to help convey what we stood for. We believed we gave our full vision and our brand identity checklist. We believed this was the most obvious brief-to-execution ever created. We believed we'd nail this in one shot… you see where this is going… we are about to pivot, and pivot hard, soon.

Figure 8.2 Phase 2 package design.

We received back multiple product packaging design concepts and very quickly realized our vision was not as crystal clear as we thought. Nor did our brief really give any kind of visual cues or aesthetics as to what a consumer would be tasting. Don't get us wrong, these designers were true experts in their domains and came forward with exactly what we requested, we just started this process too early. We took our ideation back to the drawing board and started to work simultaneously with our R&D and business development ideas to see if our juices could flow more creatively while in sync.

As we simultaneously worked on a 360° approach we realized our biggest mistake was not coming up with ideas and sketches ourselves of what resonated with us (Figure 8.2). Enter ideation phase two. Now, this is after we realized there was a direct correlation to popcorn in the making. We stuck to our keywords: quirky, bold, fun, approachable and hopeful for something 'punny'. For our quirky aspect we knew our brand images and website had to feel unique. Our product shots and photoshoots often showcase bold background images with product samples and kids toys across these hero shots. Yes, kids toys – not only did we have angry little toddlers with missing train tracks and bowling pins, we also had to Photoshop out their greasy little *Coup D'ePop* fingerprints off the toys post-shoot. Don't worry, there was a silver lining. We conducted legit focus groups on children, parents and their response to touch, taste, and feel of our product.

As we conducted these mini focus groups across multiple audiences, it became imperative to begin our brainstorming and exploring how to communicate the essence of our product, especially now that we finalized on our initial roll out of flavors.

Which keywords could we throw out to the Western masses to help define the nature of the product we were working with? Initially, it was most important to translate the Indian term 'makhana' into a relatable Western product category. Was it a puff or a seed? Would 'pop' be more appropriate to communicate its effect? Did it come from a lotus plant or a lily plant? Our drawing board showed all combinations to properly define it: lotus pops, lily puffs, lily seeds, lotus seeds, water lily pops. In deciding which product would truly do justice to its physical properties, while still conveying its similarity to popcorn, we ultimately went with 'roasted water lily seeds'. Once its physical attributes were defined, we took the initiative to deepen our exploration into finding keywords in an effort to find an all-encompassing brand name.

'Disruptor', 'revolutionary', 'market shifter', 'healthy', 'pure', 'natural', 'mass' – when exploring impactful buzzworthy words these stood out as ones to shout at the top of our lungs to create a noteworthy stir. Our quest to find an all-encompassing name to include all these words for the brand, without saying them outright, became our next goal. We knew we wanted to keep the voice of a brand that was quirky yet impactful and powerful without losing its approachability factor.

Then began the back and forth with our word play involving the word POP and how the use of the word could also associate 'popcorn' with our product. We had already established that water lily seeds having properties familiar to popcorn would be another major selling point. Appearing out of thin air as an alternative and healthier option to popcorn would be our strongest foot forward, but we would need to tread the waters carefully in our efforts to not attack the popcorn industry, but merely show that our goal was to offer a viable alternative.

Slogans like 'The Newest POP in town', 'The Best POP that's not Popcorn', 'POP it like it's hotter than POPCORN', and 'Taste the POP that's not corn' began to surface as we strategized to align our word choice with popcorn to be easily associated with it (Figure 8.3). Two reasons made this a very important message. One, relatability and familiarity which we discussed earlier. And second, while our product may actually be a seed, our goal was to position ourselves on grocery shelves next to Popcorn because of its matching snacking experience. The brand and packaging had to encompass and convey that very experience to our snacking consumer.

This is where the name Coup D'ePop emerged. It encompassed every selling point and brand trigger we strived to communicate with our consumers without losing its ability to be fun, quirky, punny – while still holding

Figure 8.3 Early mock up slogans during brainstorm sessions.

a strong narrative. The revolutionary nature of our product and our strategy to create uproar with the masses brought about the idea of a Coup of sorts, a revolution against an existing product, popcorn. A play on the word 'coup d'état' led us to 'Coup D'ePop' to communicate a name that was disruptive enough to ideally shift the snacking market from popcorn to roasted water lily seeds. With a name so impactful and a call-to-action so powerful, we were at the stage of designing our brand DNA including imaging, copy, and packaging.

In addition to dissecting our brand DNA, we knew we had to work hard on conveying the popcorn-substitute aspect. We needed to go behind the genetic make-up of popcorn compared to the genetic makeup of roasted water lily seeds to truly offer a 'better for you' approach when asking consumers to 'coup d'epop' with us. While doing this we focused on keeping our ingredients as whole and pure as possible and reviewed our popcorn competition to the healthiest offering of popcorn brands. At this point, we were at the perfect juncture to test our thesis through a market analysis and turned to BimiLeap.

BimiLeap, is an online research platform that utilizes a variation of conjoint analysis pioneered by Dr Howard Moskowitz. This process is a powerful stimulus–response experiment. The study determines what resonates in a possible consumer's mind by listing various attributes of a product, mixing them in a proprietary systematic way, presenting them to the consumer, and obtaining ratings. It simulates Nature, as a full experiment, investigating the final product of the algebra of the consumer mind. How different propositions of ideas drive a person's response. How ideas influence each other. Most importantly, the participant cannot game the system.

Using this approach, four areas of interest were studied, as listed: the event of consumption, the benefits of the food, the differentiators from other

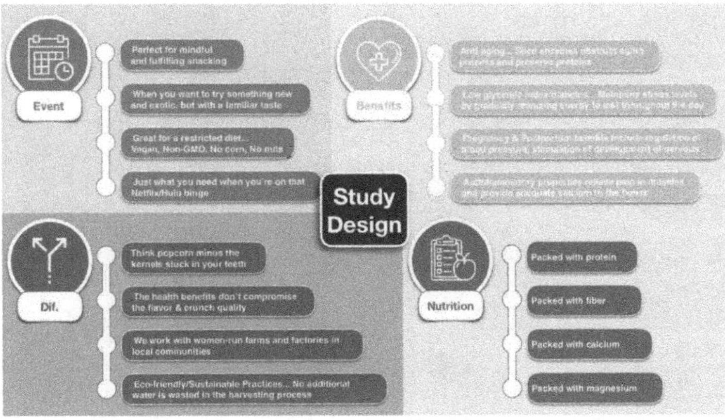

Figure 8.4 Bimileap's testing criteria for internal review.

brands, and the nutritional benefits (Figure 8.4). Each of these categories had four answers/variations.

Instead of testing Coup D'ePop, the general concept of an alternative popcorn was introduced to the respondents, since lily pops would most likely be an unfamiliar food.

The study consisted of 30 respondents varying in age (18–55 + years) and splitting near evenly between the sexes (14 male, 16 female). However, instead of demographics, segmentation based on mind-type was the main focus of this test. By analyzing the data, three segments of the population were identified. The first being individuals driven by crunch and flavor. The second segment was significantly turned off by any pregnancy benefits, but appreciated other nutritional bonuses. The third segment was interested in new and exotic flavors (Figure 8.5). Each group was close in size and similar in demographic spread, reinforcing the notion that demographics don't tell the full story. The results suggested that it was best to target segments one and three because of their higher baseline interest in an alternative popcorn and multiple overlapping drivers.

Taking into account our learnings from BimiLeap's market study, we were able to narrow down our message and target audiences. It was now time to revamp our packaging to ensure initial trial, retrial, and eventually conversion. What were the most important components to make sure this experience happened? Visual cues, relatability, and finally awareness.

Impactful visual cues prove to be eye catching to consumers in an oversaturated snacking market. How can you stand out to shoppers on fully stocked grocery shelves? Vibrant colors, block letter easy-to-read fonts, and well thought-out distribution of packaging real estate is how. When you combine this with the playful, quirky, and whimsical nature of Coup D'ePop, a personality is formed that reflects the brand voice.

Generally, bold colors signify bold statements, and we love bold. Incorporating 'pop' colors (yes, pun intended) into our packaging was imperative in our

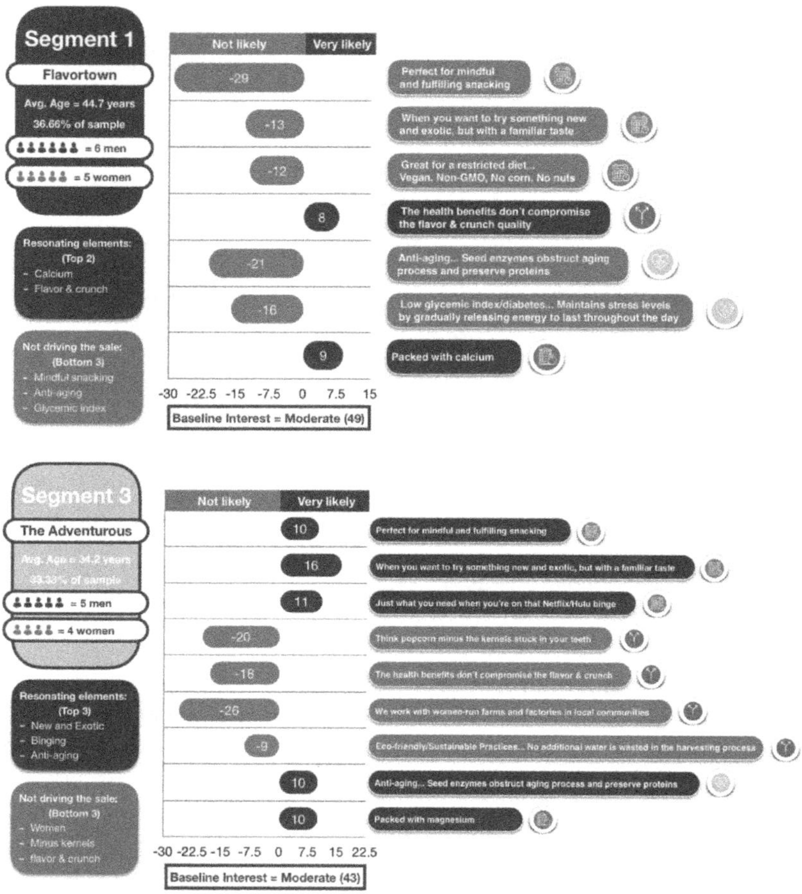

Figure 8.5 Bimileap's data review by customer segment.

branding color palette. Accompanied by an impactfully bold serif font, we were able to convey a modern straightforward and approachable voice. Lastly, the visual on packaging had to not only explain what the product was, but show the product, explain flavor profiles and highlight key call-outs that would resonate with consumers looking to fill their pantries with healthy snacks. Allocating these components to the real estate of the design of the packaging required strategizing regarding placement based on importance and size. We made sure that flavor profiles and strong visual images were at the forefront of our packaging to ensure messages of relatability and familiarity resonated with consumers (Figure 8.6).

The top feature on the packaging needed to include actual images of the product, roasted water lily seeds, in all their misshapen glory, so that consumers knew what they were sticking their hands into. Another top feature to include would be the rendering of a carnival or movie theater popcorn bag holding the water lily seeds to bring a visual experience to light – to give the

Figure 8.6 Final packaging.

impression that eating popcorn and eating water lily seeds were similar experiences. In crafting the copy to reflect familiarity amongst consumers, text phrases like 'roasted to a crunch' and 'deliciously satisfying' supported and credited the experience of our product by describing all-encompassing words to explain what the munching experience would be.

Lastly, our learnings from the BimiLeap study solidified which health factors of our product were of the utmost importance to snackers. Protein content and calories took the lead to be highlighted on the front of the packaging, while a more in-depth representation of nutritional factors in the form of icons were displayed on the back of the packaging such as 'non-GMO', 'Vegan', 'no nut', 'no corn'. We thought it beneficial to also include a lifestyle wheel revolving around water lily seeds to see how Coup D'ePop would fit into all facets of life – on the go, fitness, school, work, and home.

We have painted the whole picture around our branding and the message we wish to communicate. Simultaneously, product development proved to be the third facet of building Coup D'ePop, along with product ideation and brand ideation. When the content and brand perception has such a confident voice, it is important and our responsibility to deliver a product that lives up to those expectations and has what it takes to revolutionize how people snack.

First things first, breaking down the genetic makeup of roasted water lily seeds. Before making health claims, we, of course, delved into researching which health benefits pertained to water lily seeds, and boy, were we impressed: containing protein, calcium, potassium, and magnesium, we joke about whether these are enough '-iums' for you because, in fact, we ourselves were shocked at how much was packed into each little seed. Diving even further into our research, we uncovered many good-for-you components revealing that water lily seeds are anti-inflammatory, have anti-aging properties, provide anti-oxidants, are low in glycemic index, and aid in post-partum maternal recovery as well as childhood development. Keeping all these properties in mind, we took to our test kitchen to then give these bad boys a flavor profile that lived up to the good stuff they were made up of.

First step in recipe building: source your ingredients. Our raw material, the seed itself, is grown and harvested in Bihar, India, as you already know. It was of the utmost importance for us to partner with like-minded women-led farms in the region to ensure quality and sustainability were preserved. We were eager to establish a relationship with a partner who was willing to teach us the ins and outs of harvesting water lily seeds as well as acting as middleman to local women-run farms. It has always been a priority of ours to establish close relations to all branches of our business and be able to give back a percentage of our sales to benefit their livelihood.

Second up: test out recipes in home kitchens. Being culinary enthusiasts, this was the most fun part. Of course we were well versed on how to roast and flavor water lily seeds with passed-down Indian recipes. The real challenge was to master flavors for the Western palette with ingredients and flavor profiles easily recognized by Western consumers. Studying how popcorn is flavored with popular favorites like kettle cooked and white cheddar, we created seasoning concoctions mirroring the same flavors. We pride ourselves on using ingredients that are simple, all natural, and most importantly ones our consumers can pronounce. After numerous trials and tweaking of recipes, we had perfected two roll-out flavors we could take to market. In doing so, we had one major task ahead of us – finding the perfect fit for a co-manufacturer.

When it came to finding a co-manufacturer in America, we were very well aware of the challenges that presented themselves. First and most importantly, how were we going to find a partner that could make our final product, a product they had never tasted, heard of, or seen before. We needed to find a co-manufacturer that could integrate an entirely new product line into their existing manufacturing equation. Overcoming that hurdle would than pose more hiccups as we proceeded. Starting off in a new category with a new product line meant we needed to be realistic about our initial minimums and we needed to partner with a co-manufacturer that understood that. Lastly, it was down to us to educate on how the final product is created. Just like we needed to educate consumers about how the eating experience was similar to popcorn, we needed to educate our partners that the roasting and seasoning process for water lily seeds was similar to popcorn, and would require the same process and machinery that is associated with popcorn production.

Extensive research and cold calling led us to a local US-based, start-up focused co-manufacturer that had the capabilities to bring our ideation into fruition.

We counted ourselves lucky to find a partner that believed in our goal of revolutionizing snacking experiences, and that was enthusiastic to grow with us. To be surrounded by like-minded individuals was very important to us. As we embarked on this new journey with our new partners, we began our initial test batches and would continue to work with them through more test batches where recipes were tweaked according to their machinery as well as cook time and heat settings. Tastings led to re-tastings; and with each tasting we grew closer to a final product that neither compromised on health factors nor flavor components. Through this process, we knew we struck gold when both flavors, Sweet and Salty and Vegan White Cheddar flavored, were well within the comparable range of other healthy snacking options without taste buds suffering through an unpleasant tasting experience.

Once the recipes were perfected, and the packaging was designed, the next step involved finding the perfect packaging vendor to get our first run of product ready for sampling and tastings. The initial debate involved a discussion of whether to go with single-serve bags or larger family-size bags. A couple of factors went into our final decision. Because we wanted to give first-time consumers the best and freshest version of our product without jeopardizing the quality of our product, it became evident we needed to go with single-serve bags. We also saw an opportunity to use our single-serve bags at samplings and tastings. So, Phase 1 involved single-serve packaging, but we kept in mind that Phase 2 of production, as we grow as a company, we would include larger bags to meet distribution needs for larger grocery stores and chains. Through research and numerous attempts of outreach to potential packaging companies, we landed on one that would not only allow us to sample final packaging but would allow us to maintain low minimums at a cost-effective rate. With that, Coup D'ePop was ready to go into its first round of production with two initial flavors to introduce into the snacking market.

Looks like we're on the right track, yes? Well, sure if you think building a brand and physically coming up with a product to touch and feel is the only end goal. Yep, we're ticking all the right boxes. Let's not forget an entrepreneur's greatest acquired skill is simultaneously multitasking across infinite facets within their business.

As one would wonder, are we looking for investors? Are we undergoing investor meetings? Did we come up with pitch decks and practice our elevator pitches?

The simple answer to all of the above is yes.

Yes, we are looking for financing, but within a timeframe that feels right for us and the brand. Yes, we network constantly and have our elevator pitch ready. We have even taken part in startup conferences, shark tank-like presentations, and round table discussions, for the sake of practicing, falling more in love with our product and getting unbiased feedback. Of course each idea, product and startup needs are unique but we would like to leave you with some insight into the areas we have found most beneficial through our process.

Alright, all you budding entrepreneurs out there, this is the part where your eyes should widen and your ears should prick up. Here is our top ten 'How do I make my idea a reality, the no filter version' checklist.

1. Build an advisory board. This may seem daunting but is crucial to keep yourself in a strong place and pace. We have an advisory board consisting of two very impressive businessmen. One is an innovative food-industry pioneer focusing highly on innovation backed by market research. He has single-handedly revolutionized tomato sauce offerings, and his resumé involved working with or consulting with every food industry giant. The other holds generations of entrepreneurial experience while seeing his father create the largest eyewear force in the industry. He himself has taken on the reins as CEO and president of Brooks Brothers and offers valuable retail and industry trends from the past, present and an outward look into the future. No matter how solidified or unsolidified your advisory board is, create one and bring like-minded, passionate, experts in different domains together to discuss your business idea. Some guidelines on advisory board etiquette. Time is precious, so discuss their commitment beforehand as well as points and stake expectations.
2. Put pen to paper. You have heard this a hundred times, but it's true. When you write something on paper/or excel, your brain visually files it away for a future date. Whatever areas you need to start working on, write a plan and dates alongside it. We have done this for our go-to-market plan, digital marketing plan, financial plan, and annual forecasting, as well as creating a pitch deck. The wonderful thing about putting pen to paper is that it is a work in progress and there is no right or wrong way to start building out your thoughts.
3. Have someone valuable to you hold you accountable. We've all been there, we've all procrastinated but here is your chance to have someone hold you accountable. Whether it be your business partner, mentor, or a friend that pushes you over and above your potential. Make sure you assign this task to someone who has your growth and your best interest at heart. We recommend not using a significant other or a family member as it hits too close to home.
4. Familiarize yourself with the idea of change, regardless of how big. No idea or business that launched was ever perfect. Everything you put out there needs to grow and change as your business is always adapting and evolving. Be prepared to grow your ideas, and don't take things personally. Know that businesses that are reactive tend to succeed fastest within the industry.
5. Be vulnerable. Some may view this as a negative trait, as an entrepreneur it is not – be open minded, ask questions, listen more, understand what your potential consumer is looking for. This falls in nicely with networking with all those around you. Learn from everyone, take in what others and your customers have to share – do it all

with your business head on of course, but let people share insight with you. Let others challenge you today and save you time and money on an easy fix to your product in the future.

6. Have a plan (realistically, a minimum of three plans). When life hands you lemons... we say it's time for lemonade! You will be handed lemons constantly, your attitude and ability to think quick on your feet will help drastically. Remember to always keep your end goal in mind.
7. Accept when you need help. We can't all be masters of each and every domain, let go of that ego and ask for some help when you need it! Most startups struggle in overwhelming areas such as financials. If you need someone for small projects talk to them about coming on board for one financial quarter and again during taxes and annual forecasting. If you feel you will require someone more frequently opt for building up your team and giving stake for a CFO.
8. Know your financials, more importantly the scalability milestones. Learn to be fluid with your numbers and accept reality, learn to forecast; even if the numbers are not actual, plug in projections and assess them as you need. Know when you need investments and what that means for your business. Identify what areas you would lose or gain control over and at which milestones financially. This is scary stuff but it is also the most rewarding stuff. Have a plan.
9. Be persistent, but listen to your market first. Persistence is such an important quality as an entrepreneur, it shows dedication and grit. You want to be top of mind for all the right opportunities but remember, nobody likes a pushy salesman. Be self-aware, be able to read the room. Put yourself second. Focus on hearing what your market lacks, what it's craving for. There will be multiple times when you can leverage this information and tap into a golden opportunity.
10. Never lose your passion. Passion is sensed, it's seen, it's contagious. Your attention to detail is passion. The way you speak about your idea, your business, your team, your product, all conveys an attitude and emotion.

As we continue our journey in taking our dream forward, we would like to give you an update of where we stand. Our packaging and production prelaunch has been brought to a screeching halt given Covid-19. (Here's that curveball.) We've taken this time to continue pivoting our go-to-market plans given the expected economy we will now launch in. (Here's our pen to paper and change factor.) Through all these changes you can bet our passion still remains at an all-time high. Rest assured, Coup d'ePop is ready to create a stir at its first opportunity. As we strive to eat better together we have one request of you: join our coup d'état with Coup D'ePop!

CHAPTER 9

Assigning People to Empirically Uncovered Mind-sets: A New Horizon to Understand the Minds and Behaviors of People

ATTILA GERE*[a] AND HOWARD MOSKOWITZ[a,b]

[a] Sensory Laboratory, Faculty of Food Science, Szent István University, Villányi út 29-43, 1118 Budapest, Hungary; [b] MindCart AI, White Plains, NY, USA
*Email: gereattilaphd@gmail.com

9.1 Introduction

Mind Genomics is a widely used market research and now psychological science technique, based on experimental design, conjoint analysis, and decision theory in a noisy environment. The objective of Mind Genomics is to uncover how a person weighs the different pieces of information presented, when instructed to make a judgment of one sort or another. Mind Genomics is being widely used as of today (2020), the most popular use being consumer research to uncover the key elements that drive decisions about consumer choices of foods and services. Mind Genomics has been in continual development since the 1980s, but has seen its most recent innovations during the past 15 years.[1] Although the earliest uses clustered around topics of an applied nature, interest in decision-making has promoted the use of Mind Genomics to more than just commercial products and services. The early studies found use in science, especially food science,

Consumer-based New Product Development for the Food Industry
Edited by Sebastiano Porretta, Howard Moskowitz and Attila Gere

Published by the Royal Society of Chemistry, www.rsc.org

but have since branched out to economics,[2] chemistry,[3] social sciences,[4] and so forth.

The popularity of the Mind Genomics approach can be attributed to features that distinguish it from other methods. One is the fact that Mind Genomics has been made available in a do it yourself (DIY), format, BimiLeap®. BimiLeap® is freely available for anyone on Google Play, and now on a web browser (www.BimiLeap.com). The objective of the DIY version is to help create a generation of students who have experience with the structured thinking required when doing studies featuring experimental design. The other reasons for its popularity are the low cost, and the virtually immediate return of data, analyzed and presented in a PowerPoint presentation; the entire process of BimiLeap® lasting ~1–3 hours from the start of the set-up to the results in usable form. The real job for the user is thus to think, not to spend all the time preparing the statistical analyses.

The actual process of Mind Genomics has also been simplified, again with the vision of educating the user, while providing interesting data. The user is required to select a topic, ask four questions which 'tell a story', and then provide four alternative answers to each question. The approach requires the user to think both in a structured fashion, and yet creatively, a combination of capabilities that may be missing on the first, second and even third attempt, but a combination of capabilities that emerge for the user by the third or fourth attempt, *i.e.*, the third or fourth 'topic' investigated by the researcher using Mind Genomics.

The definition of the topic is crucial since the whole study is built around the key topic. Since the method is flexible, the researcher can choose easily a broad topic, such as innovations in the environmental sector,[5] or a more specialized topic such as new cheese formulations.[6] For this chapter, and to illustrate what we believed to be the next stage in Mind Genomics, the 'typing tool', we present the results of a BimiLeap® study on sensory aspects of meat-free food products conducted in early 2020 among 50 respondents, members of a panel, living in California, USA. It will be easy for the reader to understand our methodological focus when we present the entire study.

9.2 The Topic: Sensory Aspects of Meat-free Food Products

9.2.1 The Four Questions, Which Tell a Story About the Topic

1. How does the product look? (appearance)
2. How does the product smell? (smell)
3. How is the texture of the product? (texture)
4. How does the product taste? (taste)

As we can see from the questions above, the questions cover the standard sensory aspects of the product. The set of questions is purposely limited to four in order to promote action, rather than reflection. It is a continuing observation of the authors that when researchers and others involved in these types of projects are given enough time and flexibility of software, there is a tendency for the scope of the project to increase, so-called 'scope creep'. The study never gets done. Four questions suffice to drive the project to action, and to iterative steps.

The third step after topic and question is to create four possible answers. Within a single question one should present alternatives, but avoid directly contradictory statements. Ideally, the four answers should address different dimensions of the question, or more concretely, different possibilities about the product. The four answers need not cover the entire range of answers to the question. Four again appears to be a 'magic number' driving towards experimentation, rather than towards 'analysis (or up-front design) paralysis'.

9.2.2 The Four 'Answers' to Each 'Question'

- A) Appearance
  - A.1. Looks exactly like real meat
  - A.2. Minor differences to meat in appearance
  - A.3. Visible meat-like chunks
  - A.4. Looks processed and synthetic
- B) Smell
  - B.1. Aroma of any normal meat
  - B.2. Slight but mostly neutral aroma
  - B.3. Strong but not unpleasant aroma
  - B.4. No apparent aroma
- C) Texture
  - C.1. Similar texture to ground meat
  - C.2. Similar texture to traditional sliced meat
  - C.3. Soft, juicy and succulent texture
  - C.4. Firmer and drier texture
- D) Taste
  - D.1. Exact flavor as any meat
  - D.2. Similar flavor but obvious differences to meat
  - D.3. Major but not unpleasant flavor differences to meat
  - D.4. Contains vegetable flavors

The reader should notice that the answers to the questions are stark, simple phrases, but not single words. There is no attempt at stylistic elegance. The reason for such starkness is that at its core, Mind Genomics is an experiment, which studies how respondents graze for information in an array of messages, and uses that information to inform a decision. Thus, the answers are set up as simple, easy to read ideas, motivated by the researcher's knowledge of the respondent's task.

9.3 The Test Stimuli – Vignettes (Combinations of Answers)

The computer program, BimiLeap®, combines the answers using a 'recipe booklet', the aforementioned experimental design. The experimental design prescribes an exact set of 24 combinations of answers; these combinations are also known by the term 'vignette'. Each vignette comprises either two, three, or four answers, with each question volunteering one answer at most. The irregular size of the vignette, and the built-in structure with answers missing in some vignettes, ensures that the answers or elements in the vignettes, our 16 answers above, are statistically independent of each other, even at the level of a single respondent.

One of the novel features of Mind Genomics is that each respondent evaluates a unique set of combinations, the aforementioned vignettes. Scientists usually eliminate unwanted variability in one of two ways. The first is to suppress the noise by testing the variable in a noise-free environment, *e.g.*, eliminating any other aspects of the situation that can affect the response. The more common method to eliminate the noise is to 'average it out', testing many replicates the same stimulus, until finally the central tendency is stable. Mind Genomics goes in the opposite direction, recognizing that variation is important and may hold key information about what is being studied, rather than variation being an unwanted interloper in the project of science. Rather than working with one experimental design, prescribing the 24 combinations (our aforementioned vignettes) to be tested by dozens of respondents in the effort to 'cancel the noise', Mind Genomics creates hundreds of related experimental designs, similar 'kernels', but with different combinations. Each respondent evaluates one of these 'permuted experimental designs'.[7]

An example of a vignette appears below. Keep in mind that the vignette is constructed according to the underlying or 'kernel' experimental design, and not constructed according to what some researchers might prefer, *viz.*, 'common sense'.

A1. Looks exactly like real meat
B2. Slight but mostly neutral aroma
C3. Soft, juicy, and succulent texture
D4. Contains vegetable flavors

When we read these elements as a whole, the picture of a possible product pops up in our mind. From this point, it is easy to decide if we like the product or not. The degree of liking is usually expressed on a Likert-scale from one to nine, where one means 'I do not like it' while nine means 'I like it very much'. An advantage of the method is that the products do not need to exist, therefore there is no restriction on the researcher. The wildest, most unorthodox ideas, new-to-the-world combinations, can also be embedded in these combinations, and tested without spending time, money, and energy on the production of the products.

From the above-mentioned steps, it might seem that the design of these Mind Genomics studies must be difficult, especially the permuted designs. The answer is both yes, the basic designs and permutations are difficult, but handled ahead of time and embedded in the system, and no, the difficulties vanish when the system is set up to be done by a computer program, automatically, almost seamlessly. The BimiLeap® program runs the studies automatically, usually in a matter of an hour or two for $n = 50$ respondents, providing a full PowerPoint report. To reinforce the original motives, educating students and providing a research tool to create new knowledge, the automated analysis is accompanied by a download of the raw data, in a form ready to be reanalyzed by standard statistical programs. For education purposes, the BimiLeap® program is 'gamified', showing the distribution of strong performing elements or answers for the Total Panel, and for both the Two Mind-Set, and the Three Mind-Set clusters.

The basic design creates and delivers in 'real time' a unique set of 24 vignettes. Each vignette differs from every other vignette. In order to conform to a balanced incomplete block design (*e.g.*, all elements are presented equally), the program produces test vignettes of various sizes. As stated above, not all vignettes contain four answers. The number of answers presented range from a low of two to a high of four. The structure of the design is the same for all respondents; only the specific combinations vary. A benefit of the design strategy is that the set of combinations is 'space filling'. Rather than the meticulous analysis of a limited part of the design space, Mind Genomics looks at a large proportion of the design space, albeit with more error at any point. The strategy provides a holistic picture, with the same elements in ever-changing combinations driving the responses. Like an MRI (magnetic resonance imaging), the system explores a lot of the design space, generating a more realistic dataset, not just a manifold through the dataset, but a manifold that has to be created with a great deal of care at the start of the experiment. The Mind Genomics exploration and knowledge creation can start as easily with zero knowledge as it can with expertise.

The coding is clear and easy to convert to a standard table. Table 9.1 shows the Excel® table. A single respondent generates 24 rows of data, one row for each vignette. The respondent identification appears in the first column, and the order of trial appears in the second column. The next 16 columns are devoted to the 16 answers or elements, the number '1' showing that the answer or element appears in the vignette, the number '0' showing that the answer or element is absent from that vignette. Although the original experimental design treated the answers or elements as A1–A4, B1–B4, C1–C4, and D1–D4, respectively, the format shown in Table 9.1 prepares the data for immediate analysis.

The final column, Dep Var, is reserved for the rating scale, which is typically an anchored category scale, also known as a Likert Scale. The scale generally has an odd number of scale points. For the BimiLeap® program the researcher can choose a scale with 5, 7 or 9 scale points.

Table 9.1 The balanced incomplete block design used by BimiLeap®. Cropped data of the first panelist is presented. Test Order gives the order of presentation of the vignettes, while letters A–D denote the questions, and numbers after the letters denote the answers to the questions.

Panelist	Test Order	A1	A2	A3	A4	B1	B2	B3	B4	C1	C2	C3	C4	D1	D2	D3	D4	Dep Var
1	1	0	0	0	1	0	0	1	0	0	1	0	0	0	1	0	0	
1	2	0	1	0	0	0	1	0	0	0	0	0	1	0	1	0	0	
1	3	0	1	0	0	1	0	0	0	1	0	0	0	0	1	0	0	
1	4	0	0	0	0	1	0	0	0	0	0	1	0	0	0	0	1	
1	5	1	0	0	0	0	0	1	0	0	0	0	0	1	0	0	0	
⋮	⋮	⋮	⋮	⋮	⋮	⋮	⋮	⋮	⋮	⋮	⋮	⋮	⋮	⋮	⋮	⋮	⋮	
1	22	0	1	0	0	0	0	0	0	0	0	1	0	0	0	1	0	
1	23	0	0	1	0	0	0	0	0	0	0	0	1	0	0	0	0	
1	24	0	0	1	0	0	0	0	0	0	0	0	1	0	0	0	0	

Table 9.1 shows a cropped portion of the Excel file, presenting the combinations for vignettes 1–5 and 22–24. As noted above, each respondent evaluates a unique set of combinations, *i.e.*, a permuted version of the basic experimental design. Thus, when we deal with the 50 respondents who participated in this study, we are actually dealing with 50×24 or 1200 different vignettes. In the set of 1200 vignettes, perhaps 5–10 vignettes will be repeated. A respondent will always test 24 unique, different, and balanced combinations, the vignettes. The permutation scheme will occasionally generate two identical vignettes, albeit for different respondents.

From Table 9.1, it is clear that the design contains more 0s than 1s due to the nature of the system. A vignette can comprise at most four elements or answers from the available 16.

The Mind Genomics 'project' is to secure knowledge about the reaction to the different elements, but in a way that is more ecologically natural. Rather than presenting the respondent with one element at a time, allowing the respondent to shift her or his criteria to accommodate the nature of the element, the Mind Genomics strategy presents the respondent with these combinations of 2–4 messages, and requests an evaluation of the combination. Neither the respondent nor the researcher has time to think in the typical fashion of a questionnaire, *viz.*, asking 'what is the correct answer that I should provide?' Rather, the respondent is faced with what Harvard psychologist William James called a 'blooming, buzzing confusion'. The respondent has no clue about what is 'correct', and must simply rely upon an intuitive sense.

By analyzing the results of the 24 vignettes, we have enough input data to assess the impact of the individual answers on the final rating. The BimiLeap® program thus presents each respondent with the requisite combinations according to the experimental design, records the response (rating), makes a transformation when requested, and builds a model using ordinary least-squares (OLS) regression. The higher the positive regression coefficient, the more important the element or answer is in 'driving' the response. The negative regression coefficients show 'do not drive the response,' *i.e.*, 'irrelevant'. For further information on the data analysis of this input table, including the transformation of the Likert scale to a binary scale see Chapter 10 of this book.

The OLS regression may be done on the data from each respondent separately, or on the data from all the respondents, or even from a subset of all the respondents (*e.g.*, male). Unless otherwise specified, the data to be presented here will emerge from the analysis of respondents, one at a time; analyses that can be done easily because the underlying experimental design is implemented at the level of the respondent.

The OLS regression generates one coefficient for each answer from A1 to D4, thus creating the following set of data (Table 9.2). Underlying the table are the following preparatory steps.

The ratings are converted to a binary scale, in accordance with the practice of consumer researchers. Ratings of 1–6 are converted to 0; ratings of 7–9 are

Table 9.2 Ordinary least squares regression coefficients of the 16 answers of the Sensory aspects of meat-free food products study. The coefficients show the impact of the answers on the final rating, after the rating has been transformed to a binary scale.

Panelist	A1	A2	A3	A4	B1	B2	B3	B4	C1	C2	C3	C4	D1	D2	D3	D4
1	36	3	37	−16	−17	11	−1	−12	−57	−36	9	−33	38	49	69	11

converted to 100. To ensure some (*i.e.*, minimal) variability of the converted ratings, which are now binaries (0 or 100), a very small random number is added to each newly converted rating. Finally, the regression model is estimated. The regression is run without an additive constant, *viz.*, through the origin. The reality is that the coefficients estimated by OLS regression with an additive constant correlate highly with coefficients estimated by OLS 'through the origin', *i.e.*, without an additive constant.

Even before proceeding to deeper analyses, it is worthwhile looking at the coefficients for Panelist #1, shown in Table 9.2. A simple observation and conclusion emerge. The observation is that the coefficients differ. The conclusion is that the highest positive coefficients of each question, *i.e.*, the highest coefficient for A1–A4, the highest coefficient for B1–B4, and so forth, define the product that is expected to achieve the highest rating. Even though the respondent may not have seen this product, it is possible to define the product features that are likely to be rated the highest on the 9-point or the 0/100 binary scale. For Panelist #1, we surmise that the most liked product (in concept form), may be:

A.3. Visible meat-like chunks
B.2. Slight but mostly neutral aroma
C.3. Soft, juicy, and succulent texture
D.3. Major but not unpleasant flavor differences to meat

Let us now do the same thing, but instead of limiting ourselves to one panelist, we can estimate the coefficients, separately, of each of the 50 panelists, and put these coefficients into a matrix comprising 50 rows, one per panelist, and 16 columns, one per element or answer.

This matrix of individual coefficients enables us to look for patterns across respondents, using the coefficients as the information about the respondents. The objective is to discover groups of respondents, panelists, who show similar patterns of coefficients. These patterns are mathematically defined, so the search for patterns is done without considering the 'meaning of the patterns'. Afterwards, it is the task of the researcher to find an interpretation of the pattern.

The patterns to be discovered are called 'mind-sets'. The term 'mind-set' is defined as a group of participants who show the same or similar pattern of coefficients for a set of test elements. In other words, these individuals are similar in the way they respond to the set of test stimuli. It does not matter what the test stimuli are. The mind-sets are limited to the particular set of test stimuli. No other claim is made at this granular level of analysis. By creating such mind-sets, we can describe these groups and we can define elements, in our case product attributes, that resonate well with these people.

Mind-sets are generated by running cluster analysis on the collected individual results. Clustering is the general class of statistical procedures. The specific clustering method is left to the researcher. The results of clustering

Table 9.3 Total panel and mind-set results of the Sensory aspects of meat-free food products study. MS = mind-set, or cluster emerging from the cluster analysis.

Code	Answer to question/element in vignette	MS 1/2	MS 2/2	MS 1/3	MS 2/3	MS 3/3
A1	Looks exactly like real meat	24	9	−5	19	34
A2	Minor differences to meat in appearance	11	11	−7	13	26
A3	Visible meat-like chunks	15	7	−8	15	25
A4	Looks processed and synthetic	18	9	−10	20	30
B1	Aroma of any normal meat	14	14	28	20	−6
B2	Slight but mostly neutral aroma	13	13	19	22	−1
B3	Strong but not unpleasant aroma	19	19	27	28	1
B4	No apparent aroma	17	13	14	34	−1
C1	Similar texture to ground meat	9	10	5	28	−4
C2	Similar texture to traditional sliced meat	7	1	−1	15	−2
C3	Soft, juicy and succulent texture	14	7	6	17	10
C4	Firmer and drier texture	11	7	0	27	−1
D1	Exact same flavor as any meat	19	16	37	−3	18
D2	Similar flavor but obvious differences to meat	9	19	38	−13	17
D3	Major but not unpleasant flavor differences to meat	14	15	34	−21	30
D4	Contains vegetable flavors	17	13	30	−9	23

appear in Table 9.3, which shows the coefficients estimated by OLS regression, using the data from respondents in the different clusters. That is, the clustering program first identified the specific respondents in each cluster. The OLS program then estimated the coefficients (without the additive constant), using only the respondents belonging to a specific cluster. For example, MS1/2 is Mind-Set 1 of 2. The respondents for Mind-Set 1 and the respondents for Mind-Set 2 were identified by clustering. The OLS regression was then run twice, once for the data from each mind-set. In all, Table 9.3 shows five mind-sets, and the results of five OLS regressions.

The third and fourth columns of Table 9.3 correspond to the 1st and 2nd clusters of the 2-cluster solution, while the last three columns list the results of the 3-cluster solution.

The 2-cluster solution doesn't reveal substantial differences between the clusters. We can either accept that, or move onto a 3-cluster solution, which we did.

The 3-cluster solution shows dramatic and interpretable differences among the clusters, based upon the magnitude of the coefficients.

We change the language back to mind-sets, reserving the term cluster for the output of the program, and the term mind-set for the interpreted cluster.

We can see that MS 1/3 (Mind-Set 1 of 3) comprises individuals whose pattern of coefficients suggest that they place great emphasis on the answers of Smell and Texture (B and C).

MS 2/3 comprises individuals whose pattern of coefficients suggest that they place greater emphasis on Appearance, Smell, and Texture (A, B, and C).

MS 3/3 comprises individuals who place greater emphasis on Appearance and Taste (A and D).

It is important to keep in mind that the OLS regression and the clustering both treat all 16 elements or answers as independent, having no 'cognitive knowledge' of the meaning of the elements. All is done mechanically, despite the cognitive richness of the elements. It is only at the time of interpretation that we are pleasantly surprised by how easily we can define the key sensory aspects for the mind-sets. The cognitive richness of Mind Genomics is present all along, but is only used for interpretation. Everything else is 'objective', and done using strict mathematical criteria.

The results obtained describe the given set of respondents in an 'optimum fashion', at least for the respondents who participated in the study. That optimality is assured by the statistical process. Thus, the Mind Genomics experiment uncovers the mind-sets of the individuals who participate.

As it stands, the data and analyses cannot move further, beyond the individuals who participated. When, however, we can assign new people beyond the original experiment to one of the three mind-sets, we dramatically expand the utility of the study. We move beyond simply descriptive science to directed development and marketing on the one hand (give each person the product appropriate for that person), to empowered science on the other (understand, world-wide, the distribution of individuals in the three mind-sets, and the covariation of many other measurable behaviors with those mind-sets). The possibilities of such modeling are endless, once we are able to assign any new person to one of these three mind-sets.

Anticipating practical applications, we might say that the real power of the notion of mind-sets is its potential to divide the large world into these groups, the world of individuals who did not participate in the Mind Genomics experiment. The capability to generalize the mind-sets beyond the respondents tested is of inestimable importance in many fields. One need only think about the practical issues, ranging from marketing (*e.g.*, website development – which web design would generate more spending), to medical staff communication (with what style should we approach the patient), and finally client-directed services (what to offer to the clients).

Now that we have had a closer look at the basic data structure of BimiLeap® data, we will focus on the creation of a new tool (PVI) as a tool to drive the generalization of the Mind Genomics results, specifically the mind-sets discovered from affordable, rapid, iterative, but fundamentally small studies about how the 'mind' works in a specific topic area. In other words, how to apply the granularity of a small topic area to the entire world, and in so doing, how to create a true 'wiki' of the mind.

9.4 Available Methods and Their Applications

Prediction of unknown information from a limited amount of 'known' data is a common task in data science. The two most important groups of predictive models differ, based upon the nature of the independent variable, specifically the independent variable being quantitative *versus* the independent variable being qualitative.

Methods to relate quantitative information to quantitative predictions are commonplace. Many of the procedures fall into the general rubric of 'regression analysis'. The different variations of regression analysis will amaze. The methods that effect the predictions range from the simple and almost pedestrian method of OLS regression, to partial least-squares (PLS), decision trees, and artificial neural networks.

The aforementioned methods, ranging from OLS regression to neural networks, cannot be used in our Mind Genomics studies. The computer program, BimiLeap®, works with a limited, specific, customizable, granular set of elements, a limited number of respondents, and, from an experiment, emerges with the mind-sets of the respondents based upon the interpretable pattern of their reactions to the specific elements tested. The keywords here are the granularity of the ideas, and the limited number of respondents who must participate in the Mind Genomics study in order to let the mind-sets appear. It is the linkage of the specific elements in the Mind Genomics study with the responses of the particular individuals which allows the mind-sets to emerge. Thus, at its very core, the granularity of Mind Genomics is both what allows the powerful and meaningful mind-sets to occur, while at the same time confining the mind-sets to the respondents who participate. What is needed, therefore, is a way to assign new people to the granular mind-sets uncovered in a Mind Genomics study.

Since the aim of the process is to classify new respondents into existing mind-sets (groups), we need to create our 'assignment' models based on the existing input data (individual OLS regression coefficients as independent variables and mind-set membership as a dependent variable). Such methods belong to the category called supervised learning, where the models are trained and tested on existing data, while the created models are capable of classifying new instances.[8]

For development purposes, we introduce the classification results of six machine-learning algorithms, suitable for the modelling of the relations between the independent and dependent variables. The methods reflect different approaches to what we might call the 'generalization problem'.

9.4.1 Support Vector Machine

The support vector machine (SVM) model groups cases (in our case, the respondents) in an *N*-dimensional space by finding a hyperplane which clearly separates the groups from each other. SVM is a binary linear classifier, meaning that it always separates one group from the other(s). The plane should represent the maximum distance between the data points of the two classes, therefore future prediction of new cases becomes more robust. The name support vector comes from the points closest to the hyperplanes, *e.g.*, the points defining the distance between two groups.[9]

9.4.2 Naïve Bayes Classifier

Naïve Bayes classifier (NBC) is a probabilistic classifier method using Bayes' theorem with strong (naïve) independence assumptions between the features. NBC belongs to the simplest Bayesian network models, is highly scalable and requires a number of linear variables. Whereas most classification models in machine learning generate a numeric score for each case in the data set and assign class membership based on that numeric score, NBC assigns a so-called posterior class probability to a case. The method then uses this probability score to define the class the case belongs to.[10]

9.4.3 K-Nearest Neighbors

K-Nearest neighbors (KNN) is a non-parametric method used for the constriction of both classification and regression models. In the case of classification, KNN generates a feature space. Class memberships are determined based on the plurality vote of the neighbors of the given case. KNN typifies the so-called 'instance-based' group classification method. Class membership of a given point is determined by the class memberships of its closest neighbors, since KNN assigns weights to the neighbors based on their closeness so closer neighbors contribute more to the classification than farther ones.[11]

9.4.4 Random Forest

Random forest (RF) belongs to the family of decision trees that can also be trained as regression or classification models based on the type of response variable. An advantage of decision trees is that they provide a clear, rule-based explanation of the classification. However, if the trees are not pruned, *viz.*, allowed to grow too deep, they tend to learn irregular patterns present in the data set. The result is an over-fitted classification model which fails when new data points are introduced. The RF algorithm aims to train multiple decision trees on different parts of the same training data set to reduce variance, with a cost of a small increase of bias and loss of interpretability.[12]

9.4.5 Partial Least Squares Discriminant Analysis

Partial least squares discriminant analysis (PLS-DA) applies the same principles as its regression-variant [partial least squares regression (PLS-R)] as it relates a matrix of independent variables with a different matrix of the dependent variable. A major difference is that PLS-DA requires independent variables to be classes (qualitative values). Thus, the vector of the dependent variables is not continuous as in PLS regression. The basic idea of PLS-DA is closely connected to principal component analysis: the independent and dependent matrices are decomposed to score and loading vectors (and error matrices) with the maximization of covariance. The number of latent variables (PLS components) determines the complexity of the models.[13]

9.4.6 Linear Discriminant Analysis

Linear discriminant analysis (LDA) is frequently used to find a linear combination of features which characterizes or separates two or more classes of cases. LDA is closely related to analysis of variance and regression analysis, however, LDA uses quantitative variables to predict categorical variables. Due to the close relationship to ANOVA and regression analysis, LDA requires normally distributed data. LDA aims to minimize within-class variances while maximizing between-class variances, therefore providing clear separation of the classes.[14]

9.5 Performance of the Six Approaches for the Study Data

The same input matrix was introduced for all six classification models, which consisted of the individuals in the rows and the 16 elements in the columns. The last, 17th column contained the codes of the mind-sets previously defined by BimiLeap®. The 3-cluster solution was used here to demonstrate the performances of the prediction models, since classification of three groups creates a more complex environment for the algorithms. The data set comprised 50 rows, since 50 participants participated in the study. As we can see, the individual OLS coefficients serve as the basis of the classification.

The data set was partitioned into training (70%) and prediction (30%) sets. In order to get a more robust picture about the performances of the models, both internal and external validation has been used. Leave-one-out (LOO) cross-validation was used as internal validation and a separated data set was used as external validation. This means that 70% of the data set was used for training and LOO cross-validation and the remaining 30% was introduced as new data for prediction to the models. This way we generated two accuracy values, one for the training set and another one for the prediction set. We expect that the prediction set will achieve lower accuracy compared to that of the training set.

The accuracies obtained appear in Table 9.4. As expected, prediction sets gave lower accuracies. All six models gave acceptable performances, since they performed well above random accuracy, which is 33.3% in the case of three classes (mind-sets). RF proved to be the least accurate with 70.0%

Table 9.4 Model accuracies obtained after leave-one-out cross validation of the training set (70% split) and prediction set (30% split). The table is sorted in descending order by performance in the prediction step.

Models	Training	Prediction
PLS-DA	90.0%	83.3%
SVM	86.6%	83.0%
NBC	87.1%	80.0%
LDA	81.8%	80.0%
KNN	81.1%	76.6%
RF	71.0%	70.0%

prediction accuracy, whereas PLS discriminant analysis gave the highest accuracy for the prediction set, with 83.3%. However, SVN, NBC, and LDA also had accuracies ≥80.0%, meaning that their prediction performance is almost equally good.

The highest drops between the accuracies of the test and training sets were found in the cases of NBC and PLS discriminant analysis, whereas the smallest difference was found in the case of RFs.

9.6 Personal Viewpoint Identifier (PVI)

A different but easier and faster way of assigning new people into existing mind-sets is to extract the key elements explaining the main differences among the mind-sets. This means that we are looking for those elements which best differentiate the mind-sets. The proposed PVI is a tool that both uncovers these key elements, and creates a suitable, scalable, and affordable system to assign new individuals into the previous uncovered mind-set. The PVI defines the notion of 'distances' among the mind-sets in terms of elements. Those elements which explain the highest differences among the mind-sets are chosen as predictors.

The PVI is created using a user-friendly system, embodied in a computer program. The web-site for the program is www.PVI360.com.

The PVI begins with a link that can be sent to any respondent, world-wide. The respondent clicks on the ink, and is guided to the PVI program at www.PVI360.com. The process begins with a series of questions about who the respondent is, and so forth. These questions can be suppressed by the researcher if desired. The PVI then shows the six most discriminating elements which are presented to the participants on a simple screen (Figure 9.1), prefaced by classification questions, and followed by four additional questions that may shed light on other factors involved in mind-set membership (not shown). The researcher can suppress any of the up-front classification questions before the PVI, and chose 0–4 additional questions after the PVI.

Participants need to agree to participate and are then asked to rate the defined six elements on a binary scale. The answers on the binary scale will

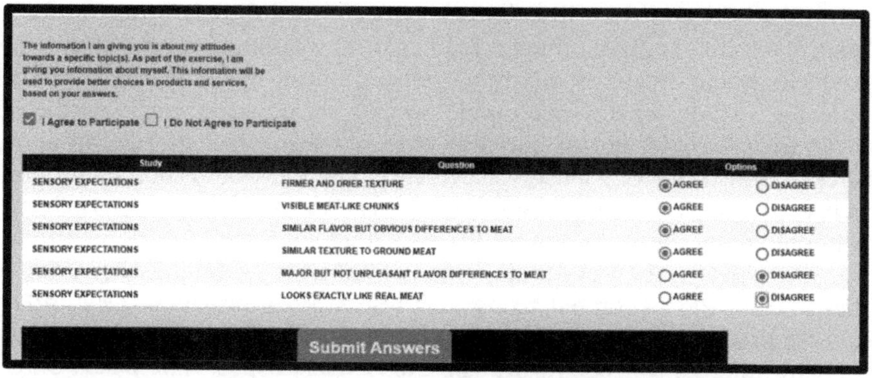

Figure 9.1 Personal viewpoint identifier for the study 'sensory expectations of meat-free products'.

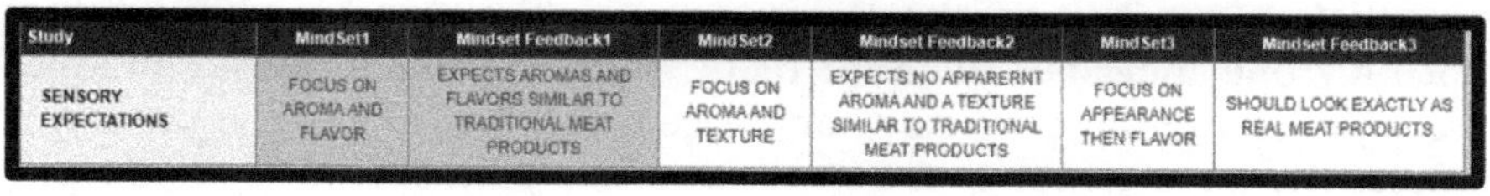

Study	MindSet1	Mindset Feedback1	MindSet2	Mindset Feedback2	MindSet3	Mindset Feedback3
SENSORY EXPECTATIONS	FOCUS ON AROMA AND FLAVOR	EXPECTS AROMAS AND FLAVORS SIMILAR TO TRADITIONAL MEAT PRODUCTS	FOCUS ON AROMA AND TEXTURE	EXPECTS NO APPARERNT AROMA AND A TEXTURE SIMILAR TO TRADITIONAL MEAT PRODUCTS	FOCUS ON APPEARANCE THEN FLAVOR	SHOULD LOOK EXACTLY AS REAL MEAT PRODUCTS

Figure 9.2 Feedback screen of the PVI. The predicted mind-set is shaded. The other mind-sets are also presented.

then decide which mind-sets they belong to with the highest probability. The binary scale is easy to use and reduces the time needed to participate.

PVI provides the final membership within 15 seconds. With the analysis, the whole process takes about 1 or 2 minutes for one PVI. The final memberships is presented to the participants along with the other possible mind-sets to provide a basis for comparison (Figure 9.2).

Not only the mind-sets, but also a short explanation of them, is presented so the user can use the information immediately. This way the PVI is suitable for situations where we are interested in the final memberships of the participants. Such situations could emerge in any circumstances: businesses who want to know the mind-sets of their clients, doctors who are interested in the ways of communicating with their patients, or even schools who want to classify their students based on their learning styles. Since the PVI is web-based, it is quite suitable for redirecting participants to personalized web pages of the same business.

9.7 A Vision of the Future – Deep Knowledge of the Mind of Millions or Billions of People

The traditional world-view has been to focus on tools to predict the membership of a new person in a given mind-set or segment. The vision of Mind Genomics is to create one, five, ten, twenty or more such segmentations, in different but related areas, such as food, create studies, create a separate PVI for each study, and then link these separate PVIs together in one exercise. Thus, if we were to have done six studies on the different topics of the 'new meat', we would create six different groups of mind-sets, one group for each topic. Each study, created in a rapid, low cost, iterative manner, would cover one of the key topics in this new area of meatless meats, *e.g.*, environmental aspects, health aspects, sensory aspects *etc.* The PVI would then incorporate each of the six PVIs, with the order of the PVI randomized for each new respondent, and with the order of the six questions randomized within the PVI, respectively. The PVI for six topics would then be 36 questions, not six questions. A PVI generated using all these studies provides the mind-set membership of the given participant for all the studies, *e.g.*, his/her mid-set of environmental aspects, health aspects, sensory aspects *etc.* Based on the pattern obtained, detailed, genome-like information is obtained, which describes the mind of the participant.

If we do not stop here but make one more step forward, it is easy to see the potential of PVI when we create several sets of BimiLeap® studies, covering not only one topic but many of them, starting from meat-free products through dairy products to fruits and vegetables. The higher the number of studies the more complete the 'genome' of the participant. By mind-typing hundreds of thousands (or even millions and tens of millions) of people based on a manageable but very large number of topics, it is possible to create a true, enlightening, and actionable 'Wiki of the mind'.

Acknowledgements

The authors acknowledge Robert Sherman for collaboration in designing the PVI. H.M. acknowledges the guidance of his mentors, Prof. Yoram Wind, and the late Prof. Paul Green, both of Wharton School of Business, University of Pennsylvania, for the early guidance and encouragement they provided in the many ongoing discussions. A.G. is grateful for the support of the Premium Postdoctoral Researcher Program of the Hungarian Academy of Sciences and the National Research, Development, and Innovation Office of Hungary (OTKA, contract No K134260). The project was supported by the European Union and co-financed by the European Social Fund (grant agreement no. EFOP-3.6.3-VEKOP-16-2017-00005).

References

1. H. R. Moskowitz, A. Gofman, J. Beckley and H. Ashman, *J. Sens. Stud.*, 2006, **21**(3), 266.
2. H. Moskowitz, S. Rappaport, D. Moskowitz, S. Porretta, B. Velema, L. Rossi and M. Atwater, Product design for bread through mind genomics and cognitive economics, in *Developing New Functional Food and Nutraceutical Products*, ed. D. Bagchi and N. P. Nair, Academic Press, ch. 14, https://doi.org/10.1016/B978-0-12-802780-6.00014-6, 2017.

3. A. Gere, R. Zemel, P. Papajorgji and H. Moskowitz, "Candy Is Dandy": The Mind of Sexuality as Suggested by a Mind Genomics Experiment, in *Sex, Smoke, and Spirits: The Role of Chemistry*, ed. B. Guthrie, J. D. Beauchamp, A. Buettner, S. Toth and M. C. Qian, ACS Publication, vol. 2019, https://doi.org/10.1021/bk-2019-1321.ch002, 2019.
4. A. Gere, P. Papajorgji, H. R. Moskowitz and V. Milutinovic, *Int. J. Polit. Activism Engagement (IJPAE)*, 2019, **6**(3), 23.
5. A. Gere, R. Zemel, P. Papajorgij, D. Radványi and H. Moskowitz, Public driven and public perceptible innovation of environmental sector, in *Innovation Strategies in Environmental Science*, ed. C. M. Galanakis, Elsevier, ch. 3, https://doi.org/10.1016/B978-0-12-817382-4.00003-4, 2019.
6. A. Gere, D. Radványi and H. Moskowitz, Consumer Perspectives About Innovations in Traditional Foods, in *Innovations in Traditional Foods*, ed. T. F. Galanakis, Elsevier, https://doi.org/10.1016/B978-0-12-814887-7.00003-4, 2019.
7. A. Gofman and H. Moskowitz, *J. Sens. Stud.*, 2010, **25**(1), 127.
8. T. Hastie, R. Tibshirani and J. Friedman, Overview of supervised learning, in *Elements of Statistical Learning. Data Mining, Inference, Prediction*, Springer, New York, NY, USA, 2001, pp. 9–43.
9. I. Zoppis, G. Mauri and R. Dondi, Kernel Methods: Support Vector Machines, in *Encyclopedia of Bioinformatics and Computational Biology*, ed. S. Ranganathan, M. Gribskov, K. Nakai and C. Schönbach, https://doi.org/10.1016/B978-0-12-809633-8.20342-7, 2019.
10. D. Berrar, Bayes' Theorem and Naive Bayes Classifier, in *Encyclopedia of Bioinformatics and Computational Biology*, ed. S. Ranganathan, M. Gribskov, K. Nakai and C. Schönbach, https://doi.org/10.1016/B978-0-12-809633-8.20473-1, 2019.
11. A. Urso, A. Fiannaca, M. La Rosa, V. Ravì and R. Rizzo, Data Mining: Prediction Methods, in *Encyclopedia of Bioinformatics and Computational Biology*, ed. S. Ranganathan, M. Gribskov, K. Nakai and C. Schönbach, https://doi.org/10.1016/B978-0-12-809633-8.20462-7, 2019.
12. M. Fratello and R. Tagliaferri, Decision Trees and Random Forests, in *Encyclopedia of Bioinformatics and Computational Biology*, ed. S. Ranganathan, M. Gribskov, K. Nakai and C. Schönbach, https://doi.org/10.1016/B978-0-12-809633-8.20337-3, 2019.
13. A. T. Badaró, C. Pasquini and D. F. Barbin, *Ref. Module Food Sci.*, 2020, **1**, DOI: 10.1016/b978-0-08-100596-5.22849-7.
14. B. K. Lavine and W. S. Rayens, Statistical Discriminant Analysis, in *Comprehensive Chemometrics*, ed. S. D. Brown, R. Tauler and B. Walczak, https://doi.org/10.1016/B978-044452701-1.00024-7, 2009.

CHAPTER 10

Systematics and Researcher Proclivities in Product Design: History, Success, and Pivot to the Future

HOWARD MOSKOWITZ,[a] DAVID MOSKOWITZ,[c] ROBERT SHERMAN,[d] ATTILA GERE,*[b] DALMA RADVANYI[e] AND RYAN ZEMEL[a]

[a] Mind Genomics Associates, Inc., White Plains, USA; [b] Szent Istvan University, Budapest, Hungary; [c] DM Analytics, Flushing, New York, USA; [d] Robert Sherman Programming, Rancho Santa Margarita, California, USA; [e] Centre for Agricultural Research, Martonvásár, Hungary

*Email: gereattilaphd@gmail.com

10.1 The Beginnings of Product Design – The World of Description

As we enter the third decade of the 21st century, we are surrounded by an abundance of tools to understand the dynamics of products, whether the dynamics concerns the actual products themselves, or the plans to develop the product, *viz.*, the prototypes or the product concepts. One might think that with this plethora of tools the success rate of food and beverage products would be greater than the dismal percentage, sometimes estimated to be as low as 10–30%, depending upon which newsletter one reads. Here is a recent quote from a newsletter from 2017.[1]

Consumer-based New Product Development for the Food Industry
Edited by Sebastiano Porretta, Howard Moskowitz and Attila Gere

Published by the Royal Society of Chemistry, www.rsc.org

'What percentage of new products fail and why?

I found these answers to 'how many new products fail?' using a quick Google search.

- According to Harvard Business School professor Clayton Christensen, each year more than 30 000 new consumer products are launched and 80% of them fail.
- 30 000 new consumer products are launched annually, 95% of them fail.
- The failure rate for new products launched in the grocery sector is 70 to 80 percent, according to Inez Blackburn of the University of Toronto.
- Ask anyone what percentage of new products fail. The usual answer is somewhere between 70–90 percent.
- It's a bit scary to hear the Nielsen statistic that more than 85% of new CPG products fail.
- 95% of new products introduced each year fail.
- Up to 80% of new product launches in the consumer packaged goods (CPG) industry fail.

Based on the above findings, I think it's safe to settle on about 80%.

The next obvious question is what causes 80% of all new products to fail. A Harvard Business Review article https://hbr.org/2011/04/why-most-product-launches-fail written by Joan Schnieder and Julie Hall states, 'The biggest problem we've encountered is lack of preparation: Companies are so focused on designing and manufacturing new products that they postpone the hard work of getting ready to market them until too late in the game.'

The reality is that the problem is not just one of faulty up-front research and design, but rather the competing needs to introduce a product that responds to changing consumer needs and wants, the competitive landscape, and finally the need for profit and contribution to keep the business going.

Going back a century to the 1920s and 1930s, and even to the 1940s and 1950s, times which now seem ancient, we find a remarkable absence of learned tomes and consultants specializing in new product development. Rather, development was left to the practitioner on the bench, a so-called bench chemist, who would explore different product prototypes based upon a 'hunch' that the consumer was 'ready' for an improved product, a simple, delicious, fruit flavored gel (*e.g.*, Jello®, born in LeRoy New York), or perhaps an angel's food cake with white frosting inside, or even chocolate cake with white frosting inside, and so forth. Absent were the ranks of experts proffering approaches, metrics for success, norms for introductions and so forth. New product introductions appeared when the bench developer felt ready, and the company decided 'Yes.'

There were beginnings of a systematized product development system, or perhaps better a systematized system for evaluating products, seemingly offered first on a commercial scale by the experts at the Arthur D. Little Company, in Acorn Park, near Cambridge, MA. Papers such as 'What makes flavor leadership' by such luminaries as Stanley Cairncross,[2] back in the 1950s, promoted a disciplined evaluation of flavor by experts. The bottom line of such work was the belief that humans could guide product design, albeit by becoming experts, and reporting back about the 'notes' of a product as it was being formulated.

In the end, however, the tools offered by the Arthur D. Little Company would evolve into the world of descriptive analysis experts, and essentially become a key too for sensory-based quality control. It was always claimed by those involved with descriptive analysis that the analysis led to product design, but the typical statement was something to the effect that *'the product developers know what to do with these mountains of descriptive results'*. What to do specifically does not seem to have been explicated, at least to the public.

One very important aspect of descriptive analysis is worth mentioning, because it includes product design. That aspect is the evolution of science from describing the components to then understanding how the components work together, and then learning how to create new and better combinations. The first third is *description*, which finds its parallel in the history of psychology in the movement called 'Structuralism',[3] whose underlying rationale is that knowing the components teaches about the system. The assertion is not the case, but that does not matter. The middle third is how something *works*, which finds its parallel in the history of psychology in the movement called 'Functionalism,' whose underlying rationale is that knowing how the components work teaches about the system, which it begins to do. The final third is *design from basics*, which represents the use of the information to create tools, which finds it use in the creation of tests to measure aspects of people's minds and proclivities.

10.2 The Contribution of Experimental Design, Scaling, and Psychophysics

In the 1950s a cadre of statisticians began their tenure at big corporations in the USA, professionals well-schooled in the methods of advanced statistics and experimental design. We will discuss their contributions below, but before doing that we should recognize the contribution of publishers who disseminated the growing knowledge of both theoretical and applied statistics. Foremost among the publishers was the John Wiley Corporation. The Wiley Interscience series on statistics was appearing, and some of the successes using experimental design were making their way into the public scientific literature. The Wiley Interscience series featured books on theory and application of advanced statistics, some geared to the mathematical statistician, others geared toward the practical statistician. At the same time, there were appearing in the literature scattered reports about statistical methods used in World War II and afterwards, methods that were practical, and easy to use. One of the most important was the Plackett and Burman Screening design,[4] used to evaluate the contribution of many independent variables, simultaneously interacting to drive a response. There would be others, but the Plackett and Burman design appears to be one of the earliest practical methods.

The statisticians who worked in the field did so without the benefit of the prolific literature that abounds today. Those early statisticians such as Dr Robert Carbonell of Standard Brands, Inc., and Mr Al May of Pillsbury, worked in what we might call 'splendid isolation'. They did not publish, although personal communication with both around 1973–1975 by author H. Moskowitz revealed their still-firm belief that experimental design would be profoundly valuable to food product design and development.

The literature in food science using experimental design was sparse. One of the earliest appears to be Joan Gordon's 1965 paper on experimental design in orange juice.[5] Appearing in the *Journal of Food Science*, the paper was published as a scientific investigation, rather than a proposal for a new, more efficient development system. That proposal would not emerge from food science, rather from experimental psychology, and specifically from the once arcane topics, scaling and psychophysics.

Psychologists have long worked with scales, allowing the person to be the measuring instrument. The highest level of such work is the incorporation of psychophysical thinking, scaling, and modeling into the processes of product design, driven by what is now a half century of effort, beginning in the late 1960s. The discipline of psychophysics is the oldest sub-discipline within the world of experimental psychology. The name gives it all away. The project of psychophysics is to relate the stimulus that we measure in the physical world to how we perceive that physical stimulus. The evolution of thinking in psychophysics instructs us about how psychophysics came to be a driving force in product design, even when not directly recognized as such.

The earliest 'belief' by psychologists was that the person could not act as a valid measuring instrument. Rather than having the person rate the product, it was considered more accurate to determine the smallest difference between two products on some dimension (*e.g.*, color, concentration) that could be detected or correctly recognized. This was so-called 'measurement by error'. This world view was cumbersome, poorly grounded intellectually, and would be abandoned by product developers due to its inability to contribute to the practical issues of product development. The world of'measuring by error' could not make any contribution to product design, except perhaps to show what could be differentiated and what could not. At a practical level it could not lead to better products. Indeed, one can read the Lemberger paper[6] on just noticeable differences in the sense of taste and come away with absolutely no idea how this arcane experiment could have anything to do with product design. One would be quite right.

Sixty-five years later, all would change. By the mid-1960s, psychophysicists at Harvard University in Cambridge, MA, at Brown University in Providence,

Rhode Island, both in the USA, and at the University of Stockholm, Sweden, were well on their way to studying the relation between chemo-sensory stimuli and the perception of tastes and smells, respectively. The focus was at first on the dynamics of perception, specifically the nature of how perceived intensity grows with changes in physical magnitude. Led by the pioneering work of S.S. Stevens at Harvard,[7] the effort captured the attention of young, 'newly-minted' PhDs, such as Linda Bartoshuk, William Cain, Herbert Meiselman, and author H. Moskowitz, respectively. Their efforts focused on how people perceived taste and then smell stimuli.

The real contributions of psychophysics were yet to come. These contributions were first the shift from aqueous solutions of single stimuli to solutions of two and more stimuli, and then then the shift from model systems to real food ingredients. A second shift occurred almost at the same time, namely the shift to systematics, to experimental designs in which various mixtures were tested in a structured way, with the goal of building a model showing how perception of an attribute varied with the separate components, and then with their interactions.

10.3 1970–1990 and the Inevitable Clash Between Sensory Analysis and Psychophysics Regarding Product Design

During the 1960s the specific issues involved in product design were ignored in many ways by sensory research, especially by those dealing with descriptive analysis on the one hand, and affective testing (hedonics) on the other. The former was interested in better language, the latter was interested in go/no decisions about the product, specifically was the product good enough to be sold to consumers.

The early 1970s saw the first coming together of psychophysics and sensory analysis, a rapprochement that would last ~2–3 years, and then dissipate. The issues dividing sensory research and psychophysics were deep however, beginning with the intellectual orientation and ending in a grab for legitimacy, an effort quickly won by the field of sensory analysis, in hindsight, to the detriment of product design. The issues came to the forefront in 1974, with the presentation of the quantitative descriptive analysis (QDA) system,[8] which was opposed by applied psychophysics.[9–11] QDA was an advance in descriptive analysis, using statistics rather than consensus, but not pointing the way to product design. Psychophysics, in contrast, pointed the way to design, and had done so since 1970, when presented at the August 1970 meeting of the International Union of Food Science and Technology (IUFOST), in Washington DC. At that meeting, H. Moskowitz presented the use of psychophysical scaling, along with the first use of the JAR (just about right) scale to optimize the grind of hamburger.[12] The approach was not accepted by the world of sensory analysis professionals, although it was accepted by commercial entities throughout the 1970s and 1980s.

As we look back 30–50 years from our focus on product design, it may be that the most important development of those early days was the clash of beliefs regarding what specific knowledge one should obtain from the consumer. Should the information be description, the 'structuralist' approach, with ever more detailed training of the panelist? Or should the information be relations between the physical stimulus and the response, the 'dynamics' of the product, without bothering to move beyond the consumer. After all, the data did suggest that in some important cases there were no strong differences between experts and untrained consumers.[13] The arguments are still with us today, at the time of writing (mid-2020). Systematic testing, emerging out of the world of psychophysics and now rejected by sensory analysis, would find another direction, which would in the world of consumer research be executed by market researchers.

10.4 Product Design, Ingrained Beliefs, and the Evolution of Product Design in Market Research

Any historical retrospective on the discipline of product design will inevitably reveal the influence of deeply held beliefs about the perceptual and cognitive abilities of people. In this short review it is only necessary to describe a few of these beliefs, most of which state what the person 'cannot do'. As a consequence of ingrained beliefs, the science and technology of product design has been significantly affected, and most generally weakened. Before we list four common but incorrect beliefs, it is worth noting that when the topic of limits to a panelist's ability is raised on an internet discussion board, far more people proffer opinions than for any other relevant topic. It appears that strongly-held beliefs about consumer abilities and limitations are a never-ending, 'evergreen' source of interesting controversies.

Belief 1. The consumer is incapable of anything more than saying 'like' or 'dislike'. This is a particularly pernicious belief, held by many professionals in the world of advertising, where the belief is that the only job of a consumer is to accept or to reject. No evidence backs up belief 1, but it is a belief that seems to circulate when people talk about the world views of advertising agencies from the 1950s to the 1990s. Everyday experience with foods will clearly dispel this pernicious and clearly incorrect belief.

Belief 2. The ordinary consumer cannot accurately perceive the attributes of a product. As a result, the developer needs to work with expert panels. This belief has been challenged,[13] but remains a point of contention. The reality is that for most attributes that are 'obvious' to a consumer, one simply needs to clarify what the attribute is. It is not that consumers are unable to perceive the attributes, but rather that they do not have a clear concept of the sensory attribute. The solution is simply to show the consumer different examples of products possessing that attribute, but not necessarily to 'train' the consumer to become an expert.

Belief 3. The consumer can only taste one or two products before losing the ability to discriminate. This third belief is particularly pernicious in the world of product design. Consumer respondents are necessary as the key instruments to measure 'acceptance'. Machines cannot measure acceptance, although they can be programmed to pick out basic flaws. We learn a great deal from testing a systematically varied set of four or more products, allowing us to develop a curve showing how liking (a subjective response) varies with physical ingredient (what is varied). If a consumer panelist can only test one or two products before so-called 'losing sensitivity', then it will be impossible to generate a liking-*versus*-ingredient curve at the individual respondent level. One consequence, discussed at length below, is that without this individual curve, we will not be able to uncover different preference patterns, sensory preference segments, within the population. Fortunately, this third belief has also gone by the wayside for many product development efforts. Yet the belief continues to rear its head from time to time, hindering the efforts of product development.

Belief 4. One could use simple linear regression to relate liking to the sensory attributes of food. In the 1980s, Howard Schutz of University of California at Davis, suggested that one could take the first step to 'optimize a product' by using the method of multiple linear regression. The independent variables were the sensory attributes, the dependent variable was 'liking'. The coefficients showed the importance of the sensory attribute as a 'driver' of liking. The idea was taken from econometrics, and cross-sectional modeling, where one uses a wide number of different observations in the same general area (*e.g.*, housing starts) to identify the key factors. The fatal flaw, however, was the failure to realize that the relation between sensory attribute and liking often reached the highest level, the optimum, in the middle range. The simple linear model was enticing but misleading.[14]

10.5 Early Efforts and Blind Alleys – Matching the Self-designed Ideal and Directional Ratings

The focus of the early 1970s was on two intellectual constructs, the first being the self-designed ideal, and the second being directional ratings. Looking back at the history of product design, it is not surprising that these two constructs were popular. In the world of descriptive analysis, it made intuitive sense that one could ask the respondent to hypothesize what the sensory profile of the best product would be. Furthermore, it also made intuitive sense that the respondent would be able to state the degree to which a product had too little *versus* too much of an attribute.

The use of self-designed ideals and directional ratings for product design was introduced to the field at the above-mentioned 1970 IUOFST Conference. The abstract by Moskowitz and Secrist dealt with the use of the ideal and the directional ratings for the evaluation of the grind of hamburger. The paper, supported by the US Army Natick Laboratories, and encouraged by

Dr John Kapsalis, sought to bring a new way of thinking to product design. The conference paper would be expanded to a full refereed paper.[12] A few years later the effort to create better products by creating formulas that delivered the sensory profile of the self-designed ideal would be presented to the scientific and business communities in another paper, with the method named 'Eclipse'.[15]

The problem with all of the 'directional' approaches was the assumption that the consumer respondent could communicate exactly what to do, albeit in the world of his or her own subjective experience. The ingrained assumption was that the problem to be solved was the transformation between ingredients and subjective responses, the 'job' of psychophysics. What the real issue was and remains is whether the consumer even knows what the 'best level' of a sensory attribute is? It appears from author H. Moskowitz's experience that respondents 'know' the best levels for visual and tactile attributes, *viz.*, appearance and texture. What consumers do not know is the best levels for more complex, emotion-laden attributes such as 'saltiness' for a chip, 'fattiness' for a meat product, or 'chocolatiness' for a chocolate candy.

It would be almost two decades before one could actually validate efforts to create products by formulation to the self-designed ideal, or to the perfect 'JAR scores'. The efforts used experimental design of products, with a program that could create the products so that they matched the sensory profile of the self-designed ideal. The results suggested that these products designed by the respondent and created by adjusting the ingredients were not the 'winning products' that one would expect. Some of the problems were that these products 'cognitively created' by consumers, were often lower in some aspects because the respondents were attempting to be 'politically correct'. Thus, foods were almost always lower in desired 'saltiness' and 'fattiness' than would be the case for optimal liking. The respondents almost always overrated positive attributes such as 'chocolate flavor' in candy.[11]

10.6 Countering the Early Beliefs – the Accepting World of Market Research in the 1970s and 1980s

The notion that systematically varying products, using consumers to evaluate these products, *viz.*, combinations of ingredients, and then creating mathematical models to optimize the acceptance was, in simple terms, totally unexpected, and often rejected. There were, however, pockets of acceptance, none from the world of the sensory analyst, but a number from the world of market research on the one hand, and product development and upper management on the other. Consumer researchers and product developers worked towards the same goal: bringing products that were highly acceptable to the market, where they could be sold. The method was less important than the objective.

By the mid to late 1970s, the efforts to create products using experimental design, systematic product development of many ingredients, could be said to be on the way to acceptance, at least among consumer researchers. Companies adopting the method included Pepsi Cola, Standard Brands, Pillsbury, and Unilever, among others. The common thread tying these efforts together was the 'blessing' of top management, and an enthusiastic group of product developers and market researchers.

The actual studies were designed by a cooperating group of R&D (product developers) and marketing research. Up to this time, the experience of most market researchers had been with testing two, perhaps three products. Yet, there was an openness to new methods, perhaps because the education of many market researchers was at business school. The focus of business school is on the business. There are those who focus on research when taking a business degree, but the overriding objective is to grow the business. In contrast, the typical effort of a scientist is to be true to one's perception of the correctness of the science, and often to disregard the business needs in favor of loyalty to one's profession. It is hard to judge which is more appropriate, but market researchers appear to have been substantially more effective as drivers of product design using science.

10.7 The Early 1980s – Individual Differences and the Emergence of Sensory Preference Segments

As we move across the broad plain of 50 years, we stop at one of the first major discoveries, individual differences. The notion of individual differences is certainly not new, having been shown a century ago in smell[16] and in taste,[17] respectively. It is easy to demonstrate these differences simply by walking around a restaurant, looking at who salts their food *versus* who puts in pepper; who adds sugar and/or milk (cream) to their coffee *versus* who does not, and so forth. The individual differences emerge as matters of taste preference, although there are the occasional individuals who avoid salt, sugar, and even milk, for health reasons.

What makes variation in sensory preferences so important is the practical implications for the developer. The corporate mandate to marketers and product developers often specifies that the product should appeal to a majority of the population, hopefully doing so dramatically more than current products against which the product in development is designed to compete. Nestlé, for example, was known to prescribe the 60/40 solution, that the new product should beat the competitor, with a margin of 60% preferring the Nestlé prototype and 40% (or fewer) preferring the competitor.[18]

The foregoing objective, called the '60/40 principle,' admirable as it is for product development, is often hard to achieve in the real world, despite attempt after attempt by often frustrated product developers. It is not that the product cannot be discovered, but rather it is product(s), not product. Different groups of the consumer population prefer different variations of the product. The Latin proverb is never as true as it is in food: 'Of taste one does not dispute'. Popular science author Malcolm Gladwell called it the search for the RIGHT PEPSI(S)®, rather than the search for one best product.[19]

Keys to the difficulties in creating a winning product for everyone and hints about the solution were already available in the literature as far back as 1965, and based on data from Engel in 1928. At the University of Stockholm, Professor Gosta Ekman of the Stockholm Psychological Laboratories had been working with a young student, Cai Ake Akesson. Akesson systematically varied the concentration of sucrose in water to create aqueous solutions of different sweetness levels.[20] When the respondents were asked to scale how much they 'liked' the taste, the respondents generated different curves, of the form shown schematically in Figure 10.1. Some respondents liked increasing concentrations of sugar water, some disliked them, but within the range of sugar water tested, there was one level for each person that achieved the highest rating. The concentration of sugar water corresponding to that highest rating defined the person. Ekman and Akesson now had a tool by which to describe the individuals. Moskowitz then used this approach, but instead of sugar water, the test stimulus was different variations of the same food. Furthermore, instead of the optimum level of the sugar concentration, the optimum level was the sensory level, as established either by an expert panel, or by the average ratings of the consumers. The result was a way to divide people into different groups, not by who they are, but by what they like most.[21]

It was the 'discovery' of sensory segments which truly increased the use of experimental design in product development. The early 1980s witnessed the confluence of psychophysics, entrenched now in the world of product

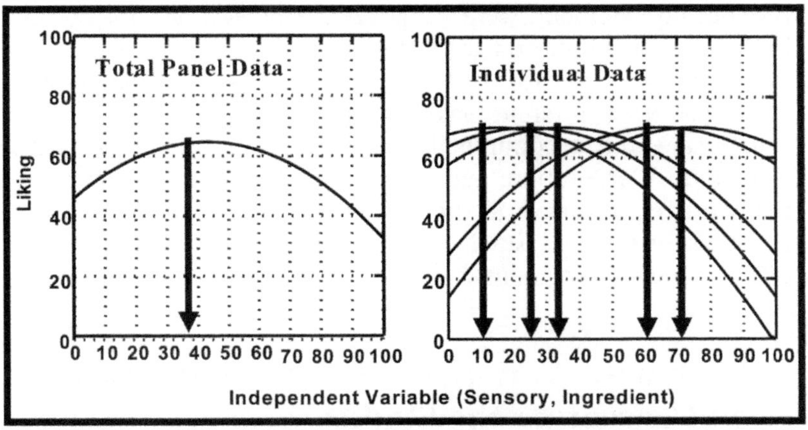

Figure 10.1 How sensory intensity 'drives' liking. The abscissa is the average rating of a product in a set on a sensory attribute. The ordinate is the individual level of liking. Each person is defined by the sensory level (on a common scale) where the liking rating for that person reaches its maximum.

development, sensory analysts focusing on descriptive analysis, and marketing researchers joining the strategy teams in companies, and tasked with the role of driving new product design.

There appears to be a 'time' when ideas come to fruition. It may be that the early 1980s was the time when experimental design was destined to begin its wider acceptance. Introduced to the corporation by some market researchers, and acquiring the blessings of market research and brand managers, experimental design was used in a variety of successful developments. The commonalities can be summarized as follows.

1. A receptive market researcher, willing to try something new, because of a sense of belonging to the company. This sense of 'safety' was important in the 1980s. It allowed the researcher to explore, something which 40 years later has disappeared almost entirely in corporations, where researchers, especially market researchers, have 'hunkered down' to maintain their jobs and to weather the seemingly endless reduction in head-count in the search for higher stock prices.
2. The increasingly inward focus of the sensory researcher, with this focus taking the shape of increased training, use of statistics to monitor panel performance, and the focus on building better testing facilities with control over the samples, the environment, and so forth. The recognition by top management of the importance of the sensory world for product description moved the sensory world away from consumers, and into experts. The consequence was the abandonment of consumer research in favor of studies with 'trained experts'.
3. Recognition by some forward-looking managers that it was possible to gain competitive advantage by testing systematically varied products. These managers, usually from marketing, marketing research and R&D product development, were recognized and rewarded for solving the business problem of how to get more sales.
4. Products emerging from these efforts were usually from large companies, whether different pickles from Vlasic Inc., tomato-based pasta sauces (Prego) from Campbell Soup Company, coffees from the Maxwell House Corporation, and so forth. Each of these companies, and many others, 'invested' in the approach, and identified both optimal products for the total panel and uncovered and optimized products for different sensory segments.

10.8 Systematic Variation – Choosing the Designs and Executing the Studies

Experimental design in the simplest instance means the evaluation of a set of prototypes which vary on one or a few aspects, whether these aspects for food and beverage be ingredients, processes, and so forth. Experimental design, the topic of this chapter, is not new. Statisticians have proclaimed the virtue of experimental design for decades. The DuPont Corporation has long promoted design of experiments (DOE), as a tool for more accurate development.

During the late 1970s and all through the 1980s, a number of companies decided to invest in experimental design for certain products, generally the products that were considered to be flagship brands. Unlike today (mid-2020), there was no sense of 'time-lines' dictating that the product had to be developed, tested, and introduced, no matter what the risk. There was no widely used process such as Stage Gate.[22] Perhaps, most important, access was far easier to high-level decision makers, whose focus was long-term growth. Today's world is quite different, as will be explicated below.

The studies typically used a fractional factorial design, and 4–6 ingredients, with 20–45 prototypes. In today's rushed and cost-conscious world, this large-scale experimental design is usually avoided, often in favor of testing 2–4 prototypes to select the winner, and calling that optimization. In the 1980s, however, the objective was to build knowledge, and often the knowledge would last for 2–3 decades, or even longer. In 2007, for instance, Michael Supran of the Campbell Soup Company stated that the data for Prego had lasted for 23 years and provided two decades of alternative formulations. The same messages were given by Dr Judy Zauenbrecher of Welches, and Dr Nicklas Ward and Mr Dennis Degeneffe of the Oscar Mayer Division of General Foods.

The key to these early studies was the coverage of a wide range of ingredients and ingredient combinations, the evaluation of prototypes by respondents who participated for 3–4 hours on 1–2 days for which they were paid, and then the creation of a model relating the ingredients to the reposes. The respondents were divided by who they were, as well as by emergent sensory preference segments determined by the pattern of liking *versus* sensory impression. Two individuals could be almost identical in age, gender, market, and product used. Yet one person might love sweet chunky products, whereas the other person might love spicy products.

Figure 10.2 presents a schematic of the response curve for two variables and the third variable being a subjective response. The left-most panel shows the how liking varies with the two ingredients, without having first

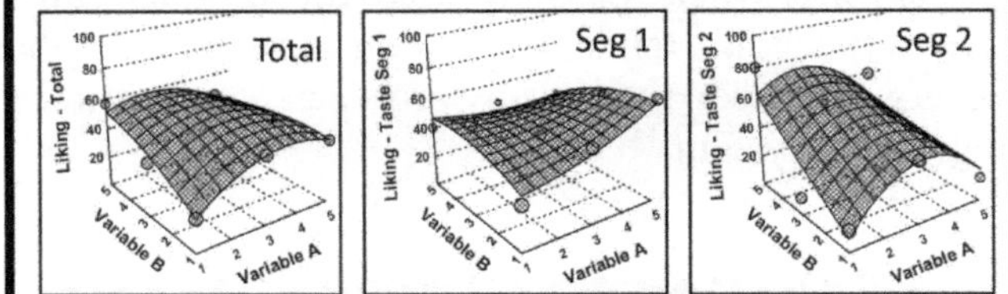

Figure 10.2 Response curve relating the ingredients and their interactions to the sensory or liking response. The surface is quadratic and schematic. The left panel shows the total, the right panels shows the two sensory preference segments, 1 and 2.

separated the sensory segments. The optimal product is in the middle, but the highest level of liking is only around 60. After the respondents are divided by the pattern of sensory attributes that they find 'best', even though they may not realize it, they fall into two sensory segments, 1 and 2, whose response curves are shown in the middle and right panels. Segment 1 likes a high level of ingredient A, and a low level of ingredient B. Segment 1's liking never goes above 60, and therefore they cannot be enticed by the formulations tested. In contrast, Segment 2's liking can go up to almost 80 with a high level of ingredient A and an intermediate level of ingredient B. None of this information would have been discovered without doing the 'experiment'. Indeed, it is not clear that the respondents would have even known their own preferences if they had not previously experienced these products.

The studies themselves were and remain straightforward to execute. They are usually executed at a central location, although with some preparation they can be done at home. The respondents are invited to participate for 1–3 sessions, depending upon the design. The session lasts 1–3 hours, depending upon the specific study. Of course, the respondents must be compensated. Over the past 50 years many companies have sprung up to recruit people to participate in these extended studies. Author H. Moskowitz pioneered this approach in 1971, almost 50 years ago, finding that the respondents were more than happy to participate when they were adequately compensated. The studies are 'choreographed' to ensure that the products are randomized, that the data are checked by an attending interviewer to maintain motivation but not suggest answers, and that the respondents wait at least 4–5 minutes between samples, sufficiently long to maintain sensitivity.[10]

The analysis of these studies generates a database allowing the researcher to identify both the taste segments, as well as the optimum product for each taste segment. Whereas the preparation for these studies seems unduly long, the execution of the studies is rapid, the analysis is even more rapid, virtually in hours, and the data are good for a decade or more. From a business point of view, the optimization approach makes a great deal of sense. The data provide a rich storehouse of intellectual property (IP), which teaches about the products, provides a reservoir of alternative products to develop over time, and enables the developer to reduce costs or change nutritional values, all the time ensuring, where possible, that the product can remain perceptually close to the 'gold standard'. In some cases, the database can even be used to 'reverse-engineer' the sensory profile of a competitor product into one's own ingredient system.[11]

10.9 Beyond the 1980s and 1990s – Process, Timelines, Budget

No paper on experimental design and the systematic creation of products can do justice to the topic without recognizing the sea changes that occurred around 1990, and persist today, 30 years later. The world of the sensory professional contracted for a while, and then began to accept the reality that it was not an insular world of expert description but had, in some ways, to participate in the world of up-front product design. In other words, the science of sensory had to become the business of sensory research, a recognition that eventually led to name changes, such as the Sensory Evaluation Division (SED) of the Institute of Food Technologists in the USA changing its name to the Sensory and Consumer Division (SCD).

In the meanwhile, deeper changes were occurring in the world of market research. The 1980s appear, in retrospect, to have been a golden age, where the consumer researcher was free to explore different approaches to solving problems, and at the same time was invited to participate in high-level decision making during the course of product design. The 1990s would see these opportunities and horizons narrow, as company after company acceded to the requests of the management to create a system of preferred (read rebate-ready) vendors. No longer were market researchers involved in high-level work. The issues now were budget, speed, and of course systematization. The process of product design was given over to lock-step processes such as Stage-Gate, so that the process was paramount, as was the timeline. It would be fair to say that the experimental designs of the 1980s were simply too much for the corporation to do, required too many resources, even though the learning, market position, and profits from the studies were continuing to 'pay out.'

The era of experimentation appears to have ended by the early 2000s, to be given over to short 'beauty contests' where the research selects the 'best' from among different prototypes, and to analytical method which attempts to substitute analytics for experimentation (*e.g.*, Mean Drop Analysis[23]). As the next section will show however, some 20–30 years after the 'lean era' began, new ideas are in play, which promise to regain the opportunity for experimentation.

10.10 Experimental Design, Rapid Testing in the Era of Available Computing

Starting around 2010 Moskowitz *et al.*[24] began to promote the idea that in light of the corporate 'pullback' on systematic designs, for whatever reason, the prudent approach to experimentation was to focus on measuring the right set of stimuli quickly, efficiently, and inexpensively. Indeed, it continued to be obvious that the answer lies in discovering the right stimulus, the right combination of ingredients. The right combination of ingredients might well be discovered with as few as 25–30 respondents, as long as they were testing the most promising combination of ingredients.

The strategy of optimization thus pivoted around 2010. Rather than finding the best product somewhere in the range of a large array of products, the strategy would be to create a feasible set of products, eight in number, systematically varied at two levels for each ingredient. The strategy generated the following benefits.

Range of variation. The eight products provided the opportunity to explore 2–7 different ingredients or process variables, each at two levels. One needed to work with at least three variables for these eight products, but with the proper 'fractionation', *i.e.*, creation of new variables from combinations of others, one could move from three variables covering the entire space (full factorial), to seven variables covering 1/16 of the space, *viz.*, 1/16 of the possible combinations. With three variables there are eight combinations. With 7 variables there are 128 combinations. Figure 10.3 shows the experimental design expanded to incorporate the ingredient name and the option for the ingredient. Thus, Variable A is flour, having two options, white and wheat, respectively. Variable G is oil, and has two options, vegetable and peanut, respectively.

The seven ingredients are generated systematically, beginning with ingredients A, B, and C, which take on two values, 1 (for the current option) and -1 (for the new option). The first three ingredients generate all 8 possible combinations of $+1$ and -1. Ingredient D is defined as the product $A \times B$. Thus, when A is $+1$ and B is $+1$, AB is $+1$. Similarly, when A is -1 and B is -1, AB is again $+1$. When A is $+1$ and B is -1, AB is -1, and so forth. This 'generating scheme' allows the experimental design of eight prototypes to cover up to seven factors or variables, each with two options.

Forced exploration. This efficient creation of eight prototypes forces the product developer to create combinations which may or may not be expected to win. It is in the nature of people to refuse to do what they think will not 'succeed'. Experimental design negates that refusal, specifying the combinations to be created. Quite often the results surprise. They may become intuitive after the experience, but the fact that some products lose when expected to win, and others win when expected to lose, is quite often the case, occasionally disconcerting, but generally cause for delight later on. Table 10.1 gives us a sense of the highest and lowest levels of product attributes.

Sensory preference segments. In the commercial world of product marketing one key objective is to get as many facings on the shelf as possible in order to present consumers with as many choices as possible from one's own product line. Marketers are famous, or perhaps infamous, for putting out line extensions simply to get shelf space. There is a down-side, however. When the line extensions perform poorly in the market, the items are pulled from the shelves, the manufacturer's reputation is damaged, and the valuable shelf space goes to a competitor with that competitor's own products. By uncovering sensory segments, in the manner of Figure 10.1, but with the new product, it becomes possible to create two or even more winning prototypes, sensorially different from each other, and strongly appealing. Table 10.2 shows the two sensory preference segments emerging from the study. It is clear from Table 10.2 that the Total panel shows no 'promising prototypes'. Yet once the sensory segments are identified based upon the pattern of acceptance scores and the relation to sensory attributes, two

#	Product Reference Number	Study Name	Product Description	Product code	Product note	Minutes Between Products	INV Var A	A Option	INV Var B	B Option	INV Var C	C Option	INV Var D A x B	D Option	INV Var E A x C	E Option	INV Var F B x C	F Option	INV Var G AxBxC	G Option
1	97	BREAD	WHEAT	111		2	IngaFLOUR	white	IngbSALT	Lo Salt	IngcSUGAR	No Sugar	IngdBUTTER	Margarine	IngeYEAST	No YEAST	IngfWATER	Tap	IngGOIL	Peanut
2	98	BREAD	RYE	212		2	IngaFLOUR	white	IngbSALT	Lo Salt	IngcSUGAR	Sugar	IngdBUTTER	Margarine	IngeYEAST	YEAST	IngfWATER	Spring	IngGOIL	Vegetable
3	99	BREAD	PUMPERNICKEL	313		2	IngaFLOUR	white	IngbSALT	Hi Salt	IngcSUGAR	No Sugar	IngdBUTTER	Butter	IngeYEAST	No YEAST	IngfWATER	Spring	IngGOIL	Vegetable
4	100	BREAD	BANANA	414		2	IngaFLOUR	white	IngbSALT	Hi Salt	IngcSUGAR	Sugar	IngdBUTTER	Butter	IngeYEAST	YEAST	IngfWATER	Tap	IngGOIL	Peanut
5	101	BREAD	FRENCH	515		2	IngaFLOUR	wheat	IngbSALT	Lo Salt	IngcSUGAR	No Sugar	IngdBUTTER	Butter	IngeYEAST	YEAST	IngfWATER	Tap	IngGOIL	Vegetable
6	102	BREAD	ITALIAN	616		2	IngaFLOUR	wheat	IngbSALT	Lo Salt	IngcSUGAR	Sugar	IngdBUTTER	Butter	IngeYEAST	No YEAST	IngfWATER	Spring	IngGOIL	Peanut
7	103	BREAD	SOURDOUGH	717		2	IngaFLOUR	wheat	IngbSALT	Hi Salt	IngcSUGAR	No Sugar	IngdBUTTER	Margarine	IngeYEAST	YEAST	IngfWATER	Spring	IngGOIL	Peanut
8	104	BREAD	CIABATTA	818		2	IngaFLOUR	wheat	IngbSALT	Hi Salt	IngcSUGAR	Sugar	IngdBUTTER	Margarine	IngeYEAST	No YEAST	IngfWATER	Tap	IngGOIL	Vegetable

Figure 10.3 The underlying experimental design showing eight bread prototypes and their ingredients.

Table 10.1 Range of products for a study on bread, showing the attribute, and across the eight bread prototypes, the key statistics: range, minimum, maximum, and median, respectively.

Total panel	Range	Min	Max	Median
Hedonics (Liking)				
DV_OVERALL_LIKING	**25**	**32**	**57**	**51**
HED_STRENGTHTAST	31	34	65	59
HED_SIZE	25	40	65	49
HED_STRENGTHAROM	21	50	71	58
HED_SALTINESS	20	43	63	51
HED_TEXTURE	19	51	70	56
HED_SWEETNESS	17	42	59	49
HED_AFTERTASTE	13	43	56	51
Image				
IMG_APPEARANCE	24	42	65	44
IMG_SIZE	22	38	60	49
IMG_SWEETNESS	20	40	60	52
IMG_TEXTURE	18	45	63	57
IMG_TASTEGRAIN	14	42	56	52
Sensory				
SEN_TEXTURE	31	29	60	47
SEN_THICKNESS	29	40	69	50
SEN_AFTERTASTE	28	42	70	53
SEN_SALTINESS	24	35	59	52
SEN_GRAINAMOUNT	22	42	64	62
SEN_SIZE	22	42	60	54
SEN_STRENGTHTAST	20	41	61	46
SEN_SWEETNESS	20	35	55	50
SEN_TEXTURE	11	50	61	54

Table 10.2 Overall liking ratings on a 0–100-point anchored scale for the total panel and for the two emergent sensory segments.

Product	Total	Segment 1	Segment 2
Promising prototypes – Segment 1			
717	57	**65**	41
818	57	62	47
Promising prototypes – Segment 2			
111	52	37	**81**
414	49	43	60
Poor performers in both segments			
515	55	57	49
313	51	54	46
212	32	28	40
616	45	52	31

prototypes emerge as quite strong (prototype 717 for Segment 1; prototype 111 for Segment 2)

Sensory–liking curve. When we plot the sensory attribute from the total panel on the abscissa, and the liking from the total panel on the ordinate, we generate a scattergram. The data may or may not look random. Quite often the data may look 'noisy' but underneath that noise there is a pattern. Figure 10.4 shows an example for the perceived amount of grain in the bread product. We can learn a great deal by running a smoothing curve through the data, *viz.*, fitting a quadratic curve. The curve suggests the underlying nature of how sensory intensity 'drives' the liking response. We learn even more when instead of plotting the data from the total panel we plot the data from the two sensory preference segments. We can get a sense of the dynamics of liking, specifically how liking changes with changes in the sensory level, and therefore whether the sensory attribute is an important driver.

Modeling to show what drives the acceptance. A key benefit of experimental design is the ability to create a simple model showing how changes in the underlying ingredients 'drive' the responses. For the efficient experimental design with eight products, each attribute evaluated by respondents can be represented by a simple model of the form:

$$\text{Rating} = k_0 + k_1(\text{A}) + k_2(\text{B}) + k_3(\text{C}) + [k_4(\text{D}) + k_5(\text{E}) + k_6(\text{F}) + k_7(\text{G})]$$

The additive constant, k_0, is the expected attribute rating when all of the variables take on the value 0, whatever 0 may be (old instead of new; absent instead of present; option 1 instead of option 2, *etc.*). That is, for dummy variable modeling, it does not matter which option takes on the value of 0, which option takes on the value 1. The additive constant is the expected rating when all options take on the value 0. All estimates are relative to the additive constant.

Table 10.3 show the coefficients. One immediately gets a sense of the true dynamics of the product. That is, one knows which ingredients appear to drive up liking, and which ingredients appear to be irrelevant, or to drive down liking. The important things about the data are the ability to isolate the key ingredients, as well as to determine a possibly stronger product than the one tested. When we pick the highest score for each of the seven ingredients, we may identify a product that has not yet been created but which has promise. For the bread being developed, we see from Table 10.3 that all ingredients should be an option, except for ingredient (variable) C and ingredient (variable) F. Changing from $C = 0$ to $C = 1$ adds 8 points to acceptance. Changing from $F = 0$ to $F = 1$ adds 7 points. Thus, the new product to be developed and tested has an expected acceptance of 62 for the total panel, but must be created and tested. The original set of eight formulations did not comprise this particular prototype: $A = 0$, $B = 0$, $C = 1$, $D = 0$, $E = 0$, $F = 1$, $G = 0$. Note further that we might get some improvement by further trying to shift E and G to option 1, but the change will not be particularly big.

Ability to satisfy multiple objectives. We often think of designed research as a method to create the 'best product' where 'best' is defined by one criterion, such as 'most liked'. Experimental design allows us to search among the various products, to define combinations that may satisfy two or more objectives, such as liking and the amount of salt (or other objective criterion), or even liking and similarity to a given in-market product, the gold standard to be duplicated (two subjective criteria). All we need is for each

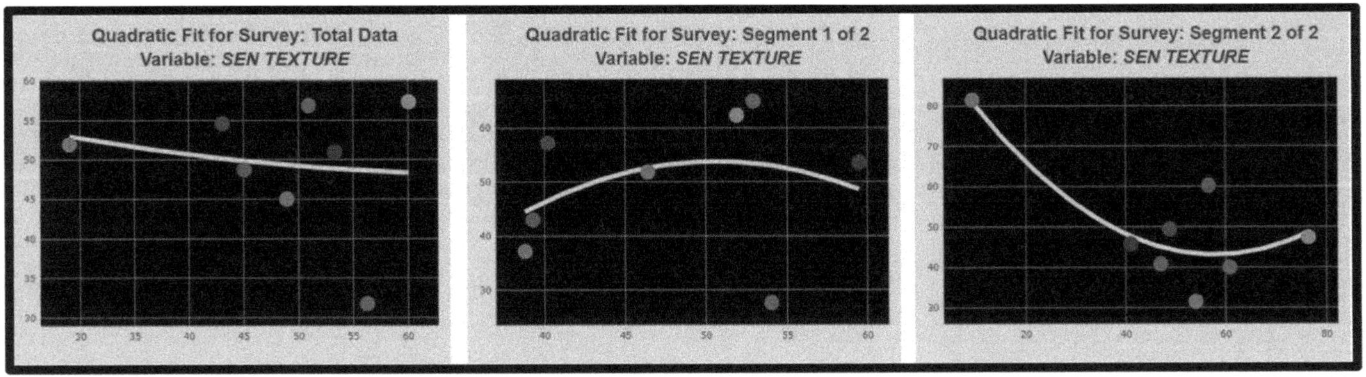

Figure 10.4 How perceived texture (soft–hard) in a bread prototype drives overall liking for the total panel and for the two sensory segments. Each circle corresponds to one of the eight bread prototypes. The curves are fitted to the data using the fitting quadratic function within the range tested.

Table 10.3 Coefficients showing how the change of ingredient from option '0' to option '1' affects the acceptance and the attribute rating for bread for the total panel.

Dep	Const	Var A	Var B	Var C	Var D A × B	Var E A × C	Var F B × C	Var G A × B × C
Additive model – total panel								
Key dependent variable	47	−8	−8	**8**	0	3	**7**	2
Hedonic attributes								
HED AFTERTASTE	66	−9	−11	1	−2	−2	−7	−5
HED SALTINESS	55	−1	5	1	−4	−11	2	−1
HED SIZE	37	−1	**11**	−6	**8**	**10**	2	4
HED STRENGTHAROM	54	**7**	−2	2	3	**8**	2	−8
HED STRENGTHTAST	54	−6	−3	−5	−11	4	3	**10**
HED SWEETNESS	52	3	−7	−3	−6	**7**	3	−2
HED TEXTURE	65	3	−6	1	−7	−7	1	−5
Image attributes								
IMG APPEARANCE	55	−12	−5	4	**6**	−6	0	−1
IMG SIZE	53	−5	1	−4	−3	**9**	1	−8
IMG STRENGTHAROM	51	2	−11	−11	−1	5	3	**9**
IMG SWEETNESS	41	4	5	−1	3	**9**	3	−3
IMG TASTEGRAIN	52	**6**	−2	3	−5	−4	0	−4
IMG TEXTURE	60	2	3	1	−7	0	−7	−7
Sensory attributes								
SEN AFTERTASTE	51	−6	−3	3	**7**	−5	4	**9**
SEN GRAINAMOUNT	50	**8**	**10**	4	−4	2	−5	0
SEN SIZE	42	**7**	0	3	−5	3	3	**7**
SEN STRENGTHTAST	56	−10	−6	4	−1	1	−5	1
SEN SWEETNESS	43	−2	−3	4	−2	**7**	**7**	1
SEN TEXTURE	49	2	−1	1	−1	2	4	5

subjective criterion to be measured on each prototype, in order to create an equation for the subjective criterion. The next task is to search through all of the possible products to discover whether there is an ingredient combination that comes close to satisfying the criterion.

Speed and discipline. As we have suggested above, creating products by experimental design forces the researcher to deal with the problems at the start of the study. Rather than 'analyzing one's way through the data', an excuse for not doing one's homework and hoping for a solution, experimental design forces the researcher to explore the different combinations. There is no room for creative 'story telling' which puts together the pieces by interpretation but leaves hanging the specific actions that the product developer must do. The results are the results. The job of the product developer is either to move with one of the products, to try one of the projected products if those are promising, or to pivot and move in another direction when none of the prototypes or none of the projected combinations have promise.

10.11 Closing Thoughts

The practices of product development inevitably follow the whims and beliefs of those in corporations, as they must. No corporation does product design and development simply for the science of discovery, at least not in the low-margin, highly competitive, and often slowly moving world of the food and beverage industries. Indeed, the statement that 'people always have to eat' has produced an industry which, up to the last half decade or so, since 2015, can be characterized as sluggish, competitive yet complacent, and reluctant to invest in new ideas.

The history of product design mirrors the history of the food industry. At the same time as we move through 2020 and into a new era of investments, it may well be that we will see new efforts in the world of product design. The old methods simply don't work, and indeed may never have worked. It was simply that the feedback loop was so long that one could substitute agile marketing for precise and targeted product development. This new decade promises the end of slow, the emergence of small and agile companies, which may or may not accept the discipline of experimental design, and the need to show real growth in order to be acquired. The acceptance of systematic product design, slimmed down, more cost effective, more agile, may well provide a forecast for the potential success of a company, which is offering products that its customers need, and not just products that the hopeful entrepreneur feels 'should be out there'. In one of Moskowitz's favorite metaphors 'the era of marketing-driven pickle dust is over; it's so yesterday'.

Acknowledgements

A.G. is grateful for the support of the Postdoctoral Researcher Program of the Hungarian Academy of Sciences and the National Research, Development and Innovation Office of Hungary (OTKA, contract No K134260). The project was supported by the European Union and co-financed by the European Social Fund (grant agreement no. EFOP-3.6.3-VEKOP-16-2017-00005).

References

1. L. Kocina, What percentage of new products fail and why? MarketSmart Newsletters, 2018, https://www.publicity.com/marketsmart-newsletters/percentage-new-products-fail (accessed 4th Jan 2021).
2. L. B. Sjostrom and S. E. Cairncross, *Food Technol.*, 1953, **7**(2), 56.
3. E. G. Boring, *History of Experimental Psychology*, Appleton Century Crofts, New York, 1929.
4. K. Vanaja and R. H. Shobha Rani, *Clin. Res. Regul. Aff.*, 2007, **24**, 1.
5. J. Gordon, *J. Food Sci.*, 1965, **30**, 903.
6. F. Lemberger, *Archiv für die Gesamte Physiologie des Menschen und der Tiere*, 1908, **123**, 293.
7. S. S. Stevens, *Psychophysics: An Introduction to its Perceptual, Neural and Social Prospects*, John Wiley, New York, 1975.
8. H. Stone, J. Sidel, S. Oliver, A. Woolsey and R. C. Singleton, *Descriptive Sensory Analysis in Practice*, 2008, **28**, 23.
9. H. R. Moskowitz, *Product Testing and Sensory Evaluation of Foods: Marketing and R&D Approaches*, Food & Nutrition Press, Inc., Westport, Connecticut, USA, 1983.
10. H. R. Moskowitz, *New Directions for Product Testing and Sensory Analysis of Foods*, Food and Nutrition Press, Westport, Connecticut, USA, 1985.
11. H. R. Moskowitz, *Food Concepts and Products: Just-in-Time Development*, Food and Nutrition Press, Westport, Connecticut, USA, 1994.
12. H. Moskowitz, *J Appl. Physiol.*, 1972, **56**, 60.
13. H. R. Moskowitz, *J. Sens. Stud.*, 1996, **11**, 19.
14. H. G. Schutz, *Food Technol.*, 1983, **37**, 36.
15. H. R. Moskowitz, D. W. Stanley and J. W. Chandler, *Can. Inst. Food Sci. Technol. J.*, 1977, **10**, 161.
16. J. H. Kenneth, *J. Exp. Psychol.*, 1928, **11**, 56.
17. R. Engel, *Arch. gesamte Psychol.*, 1928, **64**, 1.
18. E. Schmid, Personal Communication, 1982.
19. M. Gladwell, 2004, https://www.ted.com/talks/malcolm_gladwell_choice_happiness_and_spaghetti_sauce?language=en.
20. G. Ekman and C. Åkesson, *Scand. J. Psychol.*, 1965, **6**, 241.
21. H. R. Moskowitz, B. E. Jacobs and N. Lazar, *J. Food Qual.*, 1985, **8**, 169.
22. R. G. Cooper, S. J. Edgett and E. J. Kleinschmidt, *Res.-Technol. Manage.*, 2002, **45**(5), 21.
23. M. Schraidt, Appendix L: Penalty Analysis or Mean Drop Analysis, in *Just-About-Right (JAR) Scales: Design, Usage, Benefits, and Risks*, ed. L. Rothman and M. Parker, ASTM International, West Conshohocken, PA, https://doi.org/10.1520/MNL11493M, 2009.
24. H. R. Moskowitz, D. Chaiet, N. Livshits, L. Lieberman and B. Batalvi, *Empathy and Experiment: Consumer Driven Design for Sugar Reduced Snacks, Presented at Wellness 14*, Institute of Food Technologists, Chicago, IL, 2014.

CHAPTER 11

Thinking Like an Amora

HOWARD MOSKOWITZ

Mind Genomics Associates, NY, USA
Email: mjihrm@gmail.com

As I write this introduction to my new 'life's work', I would just like to confess that I am ending my 76th year, and soon starting my 78th. I will be 77 years old, G-d willing, 28 April 2021, having been born during the darkest years of our people, WWII, in New York. By way of introduction, I was born to a Chassidic family, heralding from the Zlotchover· Maggid, R. Meir of Premishlan, the Chozeh of Lublin, and R. Elimelech of Lizensk. My Zeide, Admor Israel Elimelech Moskowitz was the Zborover Rov, from the Czortkover Chassidim, author of Ner Yisroel.

If you are still reading, I beg you to go on a little journey with me. During my life, HaShem has given me the eyes to see, and the soul to help my people, the Yidn of the world. My father, Mr Moses Moskowitz, of blessed memory, guided me, telling me along the way, saying 'do something of consequence for your people.' As a little boy, when Zeide zt'l would say that in his own way, and as a young and middle age man, when Dad a'h would say that, I would nod, and hope that they would switch topics. They did. But they knew they were planting in me a seed which would take these decades to grow.

Yeshiva life, in Day Schools, first in Queens (Yeshiva Central Queens), and then in Manhattan (Yeshiva Universiy High School, MTA – Manhattan Talmudical Academy) was hard for me. I was not a good student. Certainly not a masmid. But there was more. I found Talmud hard, daunting, not relevant, not engaging. Some of the other talmidim were masmidim. I could not get the excitement of Talmud. Not then, anyway. It is hard to say those words, even now, but these days it has become important.

Consumer-based New Product Development for the Food Industry
Edited by Sebastiano Porretta, Howard Moskowitz and Attila Gere

Published by the Royal Society of Chemistry, www.rsc.org

11.1 Today's Learning Situation in Talmud... My Own Journey

I am now a much older man. I enjoy the camaraderie of studying Talmud. Yet, to be quite clear, the way I study Talmud is a much different way. I relish the arguments, the deductions, but want to contribute to it, to think of chiddushim, truly new novella, to make it mine. Perhaps I'm not really studying it like a masmid, but the truth is that it has become more deeply part of my life. I long to jump in, slowly, feel the richness, not necessarily to cover as many pages as I can, but rather enjoy the experience.

It is in that spirit that I have taken my own work, the scientific study of psychology, and put it to work studying Talmud. My approach makes the Talmud exciting to me, a non-masmid, so exciting in fact that I search when the spirit moves me to find discussions to fit a new style that I have developed, a style that I call 'Talmudic Experimentation'. In reality it is applying my own life of scientific experimentation to the Talmud, in a way which I suspect brings me far closer to the Talmud's Amoraim than I have ever been before, certainly in their spirit of learning.

11.2 Doing an Experiment with the Ideas

The notion of 'experiment' and 'Talmud' seem to be at odds with each other. We are accustomed, whether explicitly or implicitly, to hold that the word of Torah is immutable, subject to interpretation in accordance with 'revealed wisdom'. There is a 'right' way. The goal of Talmudic discourse is to find that right way, by bringing together prooftexts, using allowed principles of interpretation (*e.g.*, the 13 hermeneutic principles of Rabbi Ishmael). Once these principles have been understood, one can proceed to interpret, always being sure, however, of remaining within revealed knowledge, and accepted practice. The highest goal is for the student to bring novella that are accepted, a goal reached by few, and a goal which depresses the others, the more typical student that one encounters.

How can we impart the excitement of learning, the joy that must have been experienced by scholars during the last two millennia, scholars about whom we read, but whose very enjoyment of learning is shared by so few? Can today's technology reignite love of Torah learning, real love, and not just perfunctory expression?

> Can we make the Talmud exciting to the non-masmid, the novice amora, or perhaps even the secular student to be specific, to think in concrete terms, yet to be creative. In the Talmud creativity stops a bit with the written text, but may creep in with the discussions, and of course with the commentaries. But that's THEIR creativity. Here we want to reproduce the joy of learning, the joy of becoming creative in Torah learning. And so, the questions and the answers, modeled on Talmud study but utterly from the mind of the student, living in today's world.

What if we take the cases of the Mishnah or Gemorah, update them to day, combine the features of these cases in different ways, creating different 'what ifs', and tell different stories? But don't stop there... the effort is half done. Now continue. Present to a panel of judges these new cases, developed by the students, and have the panel of judges give a 'psak', a judgment on the cases. Finally, what happens when the student who develops these cases can see exactly how the psak, from ordinary people, is 'driven by' each idea in the case, each idea developed by the student who creates this updated Talmud case? Can we learn to create a new, engaged spirit for learning Talmud, one in which today's student is an active creator of knowledge, and by doing so becomes a deeply involved talmid, one combining the spirit of the Torah with the mind and thinking of today?

I didn't come by this new thinking accidentally. For many years, really a half century, I have been involved in business, creating new products, helping schools (even yeshivas) understand how to communicate, and being a scientist. When I helped to create the cash-back credit card in 1993 for Discover Card in Chicago, and the Oral B Mechanical Toothbrush in 1992 for Oral B Inc. in California, I used these same principles. Now I am applying them to Torah learning and Talmud.

11.3 So, What Is This Approach Really, and What Does It Deliver?

One of my favorite sections of the Talmud is *Shor Shenagach*, that ox who gored. Why? Because it's so real, so today, and has aspects of 'fun'. We all do it, in one way or another. So why not do a 'riff' on that. Not the precise topics, but the general notion of someone destroying the property of another through a third agent (the shor, *viz.*, the ox). I wanted to be both respectful of the original topic – have it relate to the sugya in the Talmud, the appropriate section, but also be creative – moving outside to explore ideas such as the behavior in the court, and whether it made any difference. To do so, I enlisted the help of a yeshiva student to 'fill in the blanks'.

We move now to a technology that I created to explore the mind. It's called Mind Genomics. Don't let the name intimidate. It is really a simple, powerful, fun approach to understand how people think about the world. Imagine we want to find out whether a person is to be let off, scot-free, or judged guilty and even fined in a case... a case recreated to follow a topic in the Talmud, but recreated by the students based on their own thinking. We follow the steps below. By the way, they are the same steps as we follow for studying to become a teacher, studying a topic in business, like what makes a good business-person, what makes two countries fight with each other, and so forth.

The ideal here is to have 2–4 students collaborate, a chavruta of students who will imagine, create, experiment, learn, and build a portfolio of Talmud-based experiments.

Table 11.1 The four new questions and the four new answers to each question.

Question 1 – What happened?	
A1	ANIMAL walked in on a tomato field and crushed it
A2	The WORKER paved a fresh road and the animal started making a real mess
A3	The ANIMAL broke the fence of the field
A4	The OWNER paved a fresh road and the animal started making a real mess
Question 2 – What got damaged?	
B1	Tomatoes got ruined to a value of $200
B2	Tomatoes got ruined to a value of $600
B3	Owner must rebuild an entire fence
B4	Animal broke the main water line... crops dried up and died
Question 3 – What happened in court?	
C1	Everyone was screaming at each other
C2	Owner attempted to kill the animal in his anger
C3	Person with the animal showed a document allowing animal walking
C4	Everyone left... not talking to each other... but mad
Question 4 – Who was doing the judging?	
D1	Judges are local farmers who know the land
D2	Judges are local magistrates (judges) who always are 'in session'
D3	Judges are regular people... local business people
D4	Judges are clergy from the religious groups in the town

Step 1. Pick the topic. Here it is a recreation of the Mishna 'Shor Shenagach', created by students.

Step 2. Ask four questions that are relevant to the topic, questions which tell a story. It's a bit of an effort, but the exercise is a powerful, effective way to teach one how to think. Table 11.1 shows us the four questions 'which tell a story' (the case), and the four answers to each question, the answers giving the specifics.

Step 3. For each question in Step 2, give four answers, each answer a phrase, not just a word.

Step 4. Ask a rating question ... for this case, what is the likely outcome. The rating question will be asked after the respondents read a 'case' as described below. The instruction is simple, and so is the rating question.

Read the sentences below. Consider them as one idea. How do you feel? 1 = Not guilty, 'pass', no problem... 5 = guilty, with a fine imposed.

11.4 Example

Take a look at Figure 11.1. Figure 11.1 has four panels, and is taken from a screen shot of the study. The computer was told that this would be 'Shor Shimmy2 BS' (short for Shor Shenagach, by Shimmy, second study, with instructions to be strict, the 'law is the law' (*viz.*, like Bais Shammai).

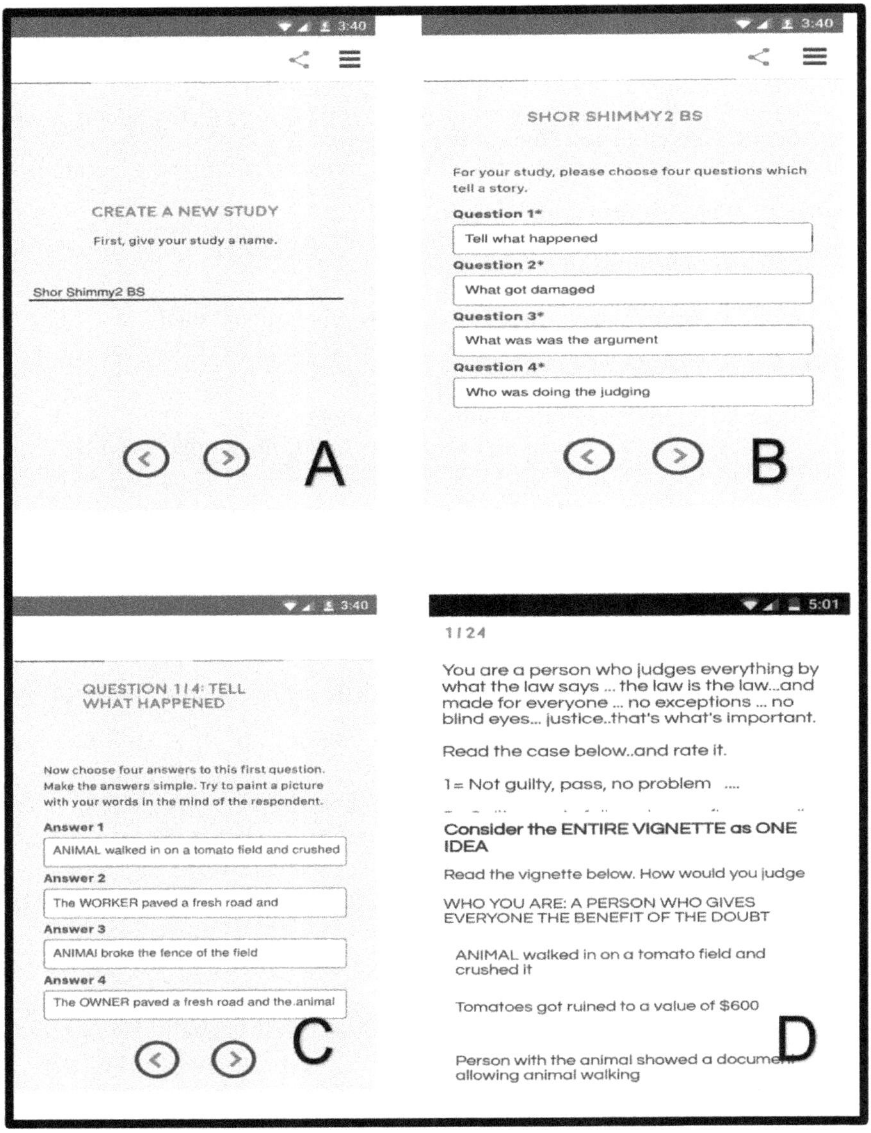

Figure 11.1 The four steps for the Mind Genomics set-up of modern-day versions of Shor Shenagach.

Panel A shows the first step. Students pick a name for the topic. For this particular study, the students (under a bit of guidance) chose the topic of 'Shor Shenegach,' the ox which 'gored'. Picking a topic is fairly easy, although one might be surprised that with a bit a pressure many students feel that they cannot 'choose' a topic. It is easy to go to a general topic in the Talmud. The term BS is 'Bais Shammai', indicating to us that when we ask the respondents to assign the rating, they will be strict, in the way Bais Shammai may have been strict.

Panel B shows the four questions. Here is the hard part. The students have to generate exactly four questions which tell a story, a task that seems easy at first, yet to the newcomer a task that often is daunting. We are not taught to think this way, in questions which tell a story. Yet that talent, that ability, will be valuable in thinking creatively and critically. The four questions are eventually developed, often taking 10–15 minutes, as the students grapple with the topic, and try to come up with a 'story in questions'. Our four questions may not be the best, but they are those chosen by the students. Note: almost everyone who starts on this process asks 'Am I doing it right? Are these the right questions and the right answers? There is no right and wrong. With practice the questions will be better, as will the answer. The chevra will become more efficient, and each student will think better.

Panel C shows the four answers to the first question, these four answers also provided by the same students. Observations over the past several years suggest that the questions are harder to develop, whereas the answers to the questions are much easier. The answers are expressed in English, in simple phrases, with few if any subordinate phrases. Table 11.1 shows the full set of questions and answers.

Pane D shows an example of an introduction to the vignette, a vignette, instructions how to rate the vignette, and one of the 24 vignettes, this vignette with only three answers, one answer each from three questions. Behind the scenes the Mind Genomics program works with a recipe book. Each respondent evaluates a unique set of 24 different vignettes, combinations of answers, combined according to pre-set recipe book. Statisticians call this an *experimental design*. The respondent has no idea that there is a recipe book dictating what is in the vignettes. Rather, the respondent simply looks at each vignette comprising 2–4 elements or answers (no more than one element per question), and then rates the vignette on the anchored 5-point scale, assuming the person to be strict and fair, without giving anyone an advantage. This is to simulate Bais Shammai. The typical time for this one rating is 3–5 seconds. The whole task takes 4–5 minutes at most. And the job is over, for everyone but the computer.

One procedure question is always asked.... What does the respondent see, and who evaluates the vignette. To answer the first question, the questions never appeared in the vignette, only the answers did. The questions are selected to motivate the answers. The Who are people from a panel (here Luc.id, a company specializing in these studies), or the respondents can be from another class, or even another yeshiva.

11.5 Making Sense of the Results – What do the Judges Say?

Each person who participates receives an email, and instructions to read the vignettes, one at a time, and make the judgment as a posek.... From

1 = nothing (let it go) to 5 (guilty, with a fine). The respondents spend about 5 seconds on each case, which is really 2–4 statements, and a rating.

For each participant, the program 'knows' the facts for each of the 24 vignettes, and the rating. Behind the scenes, the program transforms the ratings 1, 2, 3 to 0 in order to denote not guilty or really minor guilt. The program transforms the ratings 4, 5 to 100 in order to denote guilty, and even guilty with a fine.

Afterwards, and again quickly and in the background, the BimiLeap program builds 'models' showing how many points on the 0–100 point scale come from:

a. just knowing the topic (additive constant), metaphorically the ground floor of a building
b. the additional guilt (or reduction of guilt) for each element, metaphorically the height of each part of the building beyond the ground floor. There can be negative values as well, when the element reduces guilt.

So, let's look at what happened. First, we sent out 60 invitations to people, and told them that they were to judge each of the 24 cases, comprising 2–4 answers. They were to feel that they should judge fairly but to give everyone a break if possible. They were to be lenient. We tried to simulate judges who would be told to judge, but be like Bais Hillel. Another 30 respondents, totally separate participated, testing the same type of vignettes, with the exact same instructions, except told to be honest, fair, but strict. The law is the law. This was to simulate Bais Shammai. All respondents were unknown, provided by a company (Luc.id), which specializes in these types of studies.

Table 11.2 collapses the entire learning into one table, looking at the total panel. It is from the data in Table 11.2 that the student gets an idea of 'what people say' to different aspects. The data from Table 11.2 can be discussed in detail. What is nice about the discussion is that it emerges from the student's own data and is 'real'. There is no right, and no wrong. The student must think critically, both when designing/executing the 'experiment', and when interpreting what real people say.

Additive constant = likelihood of saying 'guilty + fine' in the absence of elements. The additive constant is a strictly theoretical construct, but very instructive as a baseline.

Bais Hillil actually shows a higher constant than does Bais Shammai (50 *versus* 42), or at least our instructions to simulate Bais Hillel *versus* Bais Shammai. That's a surprise. We thought it would be the other way around. Note that the respondents who participated in the study (invited by Luc.id) have no idea of our expectations. We night have expected the Shammai instructions to mean a greater likelihood to judge 'guilty' and thus a higher additive constant. What does this mean? This is a topic of meaningful and interesting discussion for to the student, armed with facts, and indeed the student's own facts!

Table 11.2 The actual coefficients for the 16 elements in the updated case of Shor Shenagach.

Group (binary ratings)		Like Bais Hillel	Like Bais Shammai
Base size		60	30
Additive constant		50	42
Question A: Tell what happened?			
A1	ANIMAL walked in on a tomato field and crushed it	−1	−3
A4	The OWNER paved a fresh road and the animal started making a real mess	−2	1
A2	The WORKER paved a fresh road and the animal started making a real mess	−6	**8**
A3	ANIMAL broke the fence of the field	−8	−4
Question B: What got damaged?			
B2	Tomatoes got ruined to a value of $600	2	−10
B4	Animal broke the main water line... crops dried up and died	1	−4
B1	Tomatoes got ruined to a value of $200	−2	−9
B3	Owner must rebuild an entire fence	−10	0
Question C: What was the argument?			
C1	Everyone was screaming at each other	7	−5
C3	Person with the animal showed a document allowing animal walking	7	−9
C4	Everyone left... not talking to each other... but mad	7	−9
C2	Owner attempted to kill the animal in his anger	−1	0
Question D: Who was doing the judging?			
D1	Judges are local farmers who know the land	−3	−2
D3	Judges are regular people... local business people	−4	−2
D4	Judges are clergy from the religious groups in the town	−4	12
D2	Judges are local magistrates (judges) who always are 'in session'	**−10**	**3**

Now let's look at the individual elements, primarily for those respondents instructed to think in the way we believe Bais Hillel might think. To make the analysis easier, we have sorted the elements within each question from highest to lowest, *i.e.*, from guilty to innocent.

Let us first plot the data, using the coefficients. Remember that the coefficients tell us about guilty (high coefficients) *versus* innocent (low coefficients). Our 'experiment' allows the students to put 'numbers' on the severity of the action. When we do this, we find, to our surprise, that Bais Hillel and Bais Shammai (or at least our instructions) lead to radically different patterns. That which Hillel finds to lead to a verdict of 'not guilty' (low numbers on the *x*-axis, the abscissa) corresponds to what Bais Shammai finds to lead to a verdict of 'guilty' (high numbers on the *y*-axis, the ordinate). This is a topic for discussion by the students (Figure 11.2). What does it mean? Or more realistically, what could it suggest?

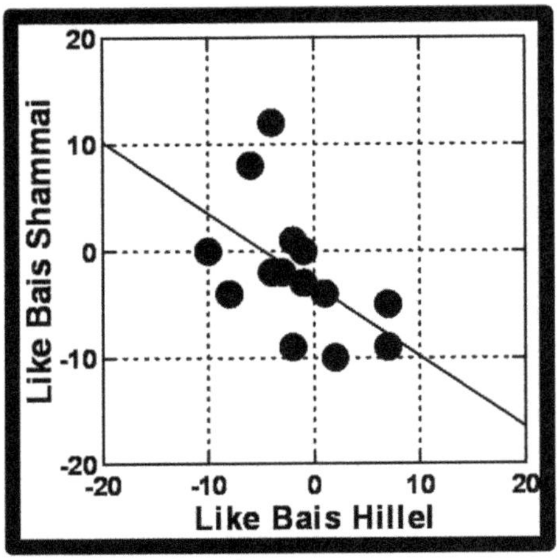

Figure 11.2 Plot of the coefficients for the 16 elements for Bais Hillel *versus* Bais Shammai. The values of the coordinates range from innocent (low, negative values) to guilty, and with an added fine (high, positive values).

The exact coefficients appear in Table 11.2. It is not important to review the coefficients here, but simply important to note that these are real results, from real people. The students who are interested in learning may want to ask: 'Why are these differences emerging the way they are?' There are no simple answers. With the experiment in hand, the student now has a sense of how different factors in the case, and even different instructions to the judge can affect the outcome, the judgment of a case. The student is forever enriched in the consideration of Talmudic cases and has a tool to expand the knowledge of the case so that it applies to today's world.

11.6 Thinking Differently about the Same Evidence

We return now to our first experiment, where we instructed our respondents to be like Bais Hillel, 'lenient'. We have worked with 60 people. A lot of times we 'know' somehow that people are different from each other but we cannot exactly define the nature of the differences. Mind Genomics helps us learn about the mind of these people who are reading the case. We divide the 60 participants into two groups, not based on who they are (we know their age and gender), but instead based upon how they 'vote'. We use the 16 coefficients for each respondent, putting the 60 respondents into two groups. The number of respondents is 32 *versus* 28, about the same number. Yet, their reactions could not be more different from each other.

Here then are people, faced with the same evidence, and the same instructions, who show radically different reactions to the same evidence. Table 11.3 compares these two mind-sets, individuals presented with the same instructions and information, yet who judge the case quite differently.

Table 11.3 The coefficients for the 16 elements (Bais Hillel group), showing the total panel, and the results from two mind-sets within the total panel.

Group (binary ratings) – when the respondents were told to judge 'leniently' (Bais Hillel)	Total	MS1	MS2
Base size	60	32	28
Additive constant	50	53	48
Elements driving 'guilty + fine' for mind-Set 1 (behavior)			
C3 Person with the animal showed a document allowing animal walking	7	**30**	−19
C1 Everyone was screaming at each other	7	**19**	−7
C4 Everyone left... not talking to each other... but mad	7	**17**	−6
Elements driving 'guilty + fine' for mind-Set 2 (judges)			
D1 Judges are local farmers who know the land	−3	−28	**27**
D4 Judges are clergy from the religious groups in the town	−4	−31	**26**
D3 Judges are regular people... local business people	−4	−27	**22**
D2 Judges are local magistrates (judges) who always are 'in session'	−10	−38	**21**
Elements which do not drive 'guilty + fine' for either mind-set			
B2 Tomatoes got ruined to a value of $600	2	3	0
B4 Animal broke the main water line... crops dried up and died	1	6	−4
A1 ANIMAL walked in on a tomato field and crushed it	−1	3	−5
B1 Tomatoes got ruined to a value of $200	−2	3	−7
A4 The OWNER paved a fresh road and the animal started making a real mess	−2	1	−7
C2 Owner attempted to kill the animal in his anger	−1	7	−9
A2 The WORKER paved a fresh road and the animal started making a real mess	−6	−4	−9
B3 Owner must rebuild an entire fence	−10	−7	−14
A3 ANIMAL broke the fence of the field	−8	0	−16

Mind-set 1 – wants order and behavior, whether from the animal or from the litigants. Even when the animal can 'legally walk' the animal is 'wrong' for destroying property.

Mind-set 2 – does not pay attention to the information in the case, but only to who the judge is. Mind-set 2 believes that local farmers acting as judges in this case of damage will be most likely to rule 'guilty with fine'. In contrast, mind-set 2 believes that local magistrates will also rule 'guilty with fine', but less so.

It is clear from this type of thinking that the student gets another dimension of thinking about the law – not just what the law is and from where, but the different ways that the law can be considered.

11.7 Going Forward – What Does This Mean for the Study of Talmud?

We are taught that there are different ways to study Torah generally, and Talmud specifically. In the 21st century of the common era, the notion of studying Talmud requires moving beyond the simple, limited, strictures of the Talmud, where one can remember, recite, and come up with answers. The requirement is to be able to think creatively, to add, and not just to repeat and recite. Modern society requires everyone, talmidim and non-talmidim alike, to be able to think creatively and critically, to adjust.

We feel that there is room in the house of Torah for critical exploration, for looking at Talmudic topics from a new perspective, one which is fun, which is creative, and indeed which allows each talmid to create his or her own future, by adding to the scope of Torah. The approach of Mind Genomics may provide just such a new direction, consistent with Torah values and with the spirit of the day.

I would like to end with one scene, that of the group of students sitting around, after they have done one of these 'Talmud Experiments'. The topic is Talmud, but the thinking is that of the students, going in their own direction. Imagine the pride the student(s) have showing the results to their teacher, results which relate to the Talmud, but which incorporate the very thinking and soul of the student(s). At once we have elevated the student(s) to a new position in the Chain of Torah. These are students, but also contributors in their own way to the Chain of Torah. Students who might not have been excited about Torah, about Talmud, now can approach the holy subjects with more confidence, and in their own ways receive the Torah anew in a way consistent with them and with their higher calling.

Subject Index